M. M. Mirzaolimov
A.A. Kobulov
K.A. Niezov N.U.

MYSTERIES OF THE HUMAN BRAIN

M. M. Mirzaolimov
A.A. Kobulov
K.A. Niezov N.U.

MYSTERIES OF THE HUMAN BRAIN

ScienciaScripts

Cover image: www.ingimage.com

This book is a translation from the original published under ISBN 978-620-7-65118-4.

Publisher:
Sciencia Scripts
is a trademark of
Dodo Books Indian Ocean Ltd. and OmniScriptum S.R.L publishing group

120 High Road, East Finchley, London, N2 9ED, United Kingdom
Str. Armeneasca 28/1, office 1, Chisinau MD-2012, Republic of Moldova, Europe
Printed at: see last page
ISBN: 978-620-7-71982-2

MINISTRY OF HIGHER EDUCATION, SCIENCE AND INNOVATIONS OF THE REPUBLIC OF UZBEKISTAN

NAMANGAN STATE UNIVERSITY

MIRZAOLIMOV MIRZOKHID MIRZAVALIEVICH
NIYOZOV QAHRAMON ADASHALIEVICH
KHODJAEVA NAIMAKHON UMIRZA-KOVNA
KOBULOV ABDULVOKHID ABDURASHIDOVICH

THE MYSTERIES OF THE HUMAN BRAIN

Namangan - 2024

UDC: *539.1.047*

Secrets of the human brain. Mirzaolimov M.M., Niyozov K.A., Khodjaeva N.U., Kobulov A.A., 2024. Izdvo. LAP Lambert Academic Publishing

Reviewers: **Ph.D. in Biological Sciences**
M.M.Mamazhanov
Doctor of Biological Sciences
A.E. Zainabidinov

Chapter

The human brain is one of the greatest mysteries of the universe. How many more unsolved mysteries are associated with brain research?

You want to know:

- where memories are "stored";

- why we have prophetic dreams;

- with what psychological techniques we're being manipulated;

- is it possible to conquer fear and find happiness?

If so, let's embark together on a fascinating and mysterious journey into the depths of the human mind to discover the secrets of fear and intelligence.

From the author

CONTENTS

INTRODUCTION

For being able to read and understand these pages, you have the gray matter in your skull to thank first and foremost. The human brain is an amazing, incredible tool for understanding the world and itself. It can't be compared to any of NASA's most advanced computers. The brain with incomprehensible speed simultaneously processes all information coming from the organs of sight, smell, hearing and touch, coordinates the actions of arms, legs, head and spine, controls every movement, turns on memory, allows you to understand and reproduce speech, is responsible for feelings, desires and emotions ... In addition, some particularly gifted individuals it allows you to think in abstract categories. In general, the brain simultaneously regulates both the vital functions of our mortal body and the rise of the immortal spirit.

Although man is the crown of creation, it is possible to speak about the presence of the brain in all vertebrates, starting with fish (in insects "brain" is a cluster of nerve fibers or ganglia). Moreover, if in humans the ratio of brain mass to body weight on average is 2-2.3 percent, then in the same primates this ratio ranges from 0.8 to 1.7, and in dolphins and killer whales is about 1.8 percent. Of course, in terms of weight, humans significantly lose to these last representatives of fauna (the average adult human brain weighs about one and a half kilograms), but we can be proud of the fact that only the human brain has such a function as thinking.

CHAPTER I. STUDYING THE STRUCTURE OF THE BRAIN: ANCIENT RESEARCH AND SURPRISING DISCOVERIES OF THE PRESENT DAY. THE BRAIN AS A COMPUTER

"Mission Control Center"

It is quite logical that the "control center" of the body and mind should be reliably protected, because any serious damage to the brain relegates a person to the category of "vegetables". In such a vegetative state, the personality stops reacting to external stimuli and loses the ability to act meaningfully. Therefore, as the first level of brain protection, nature used a case - the skull (the spine protects the spinal cord). In addition, to shield the brain matter from the pressure of bone walls, it is protected by three shells of connective tissue: hard (outer), vascular, or spider web (middle), and soft (inner). The space between them is filled with cerebrospinal fluid - liquor (from Latin liquor), similar in composition to blood plasma. Nature as if built a multi-layered airbag. And interstitial fluid in addition to the function of safety performs another one - supplies the brain with nutrients. In the brain during the day and night comes up to 20 percent of the total blood flow, carrying oxygen to the tissues of the body. The peculiarity of the cerebral circulation is such that several special membranes form the blood-brain barrier, limiting the permeability of blood vessel walls and restraining the penetration of infections and most of the various chemical compounds from the blood into the substance of the brain.

The brain is supplied with blood by three main arteries - the two internal carotid arteries (a. carotis interna) and the main artery (a. basilaris). Scientists have found that if the activity of neurons in a certain part of the brain increases significantly, the blood supply to that area increases immediately.

The deep structures of the brain

What is the structure of our brain? We will not delve into the subtleties of anatomy and physiology, because in order to understand the basic principles of brain function, it is not necessary to know where exactly the mastoid bodies or the lenticular nucleus are located and what functions they perform. In order not to get confused in terms, it is enough to remember that the brain can be conditionally divided into three main blocks: forebrain, cerebellum and brain stem.

If you were to look into a transparent skull (the case of a corpse in the morgue is less optimistic), you would first of all see two **large hemispheres** covering the deepest formations of the brain. The left and right hemispheres are the largest, most developed and most important part of the central nervous system. They are separated from each other by a deep longitudinal slit, which continues up to the corpus callosum - a bundle of nerve fibers, which is a kind of bridge connecting the two hemispheres and providing a continuous exchange of information.

The longitudinal slit ends at the transverse slit of the large brain, which separates the hemispheres from the **cerebellum** (remember, it is at the back: first the brain, followed by the cerebellum), the surface of which is covered with a network of more delicate and fine furrows and gyrus than the cerebral cortex. The cerebellum is responsible for the formation of motor skills and coordinates the actions of various muscle groups when performing stereotypical behavioral acts (such as eating, brushing teeth); it is also involved in maintaining balance.

In addition to the hemispheres, which make up about 70 percent of the brain's total volume, the thalamus (or visual center), hypothalamus (which produces neurosecretin to regulate the pituitary gland), and pituitary gland (one of the major neuroendocrine glands) are also located in the anterior region (below the hemispheres, roughly in the middle).

Beneath the large hemispheres is the brain **stem, which consists of the** midbrain (where the subcortical centers of hearing and vision are located), the pontine and medulla oblongata (where the centers regulating breathing, heart activity and blood pressure, stomach and intestinal peristalsis are located), which smoothly passes into the spinal cord. The nerves that branch off the brain stem and spinal cord collect information from internal and external receptors and send signals to muscles and glands.

All of the above brain structures are connected by special conductive structures - nerve fibers.

Large hemispheres

Let's return to the largest part of the brain - the large hemispheres. Normally, the right and left hemispheres are symmetrical and connected by the corpus callosum, and at about the level of the brain stem, the bundles of nerve fibers connecting each of the large hemispheres with the cerebellar hemispheres cross. Thus, the left side of the body is controlled by the right hemisphere and the right side by the left hemisphere.

It is generally believed that the dominant (usually left) hemisphere is more responsible for speech and accompanying gestures, as well as for other behavioral functions, while the right hemisphere analyzes spatial and temporal parameters of the surrounding space. For example, when we listen to music, the right hemisphere is more activated, while the left hemisphere is responsible for its creation.

Shortly before his sixtieth birthday, French composer Maurice Ravel was involved in a car accident that severely damaged the left hemisphere of his brain. In describing his recovery, his doctor observed a strange effect: although Ravel enjoyed listening to music, attending concerts, and making informed and intelligent criticisms, "he was never again able to compose - to put down on paper what was playing in his head," or to read sheet music or play the piano.

Interestingly, women tend to have a better developed left hemisphere, while men have a better developed right hemisphere. With all sorts of gender-awareness, psychologists are forced to admit that women are "on average" inferior to men in terms of logical, mathematical and spatial abilities, but superior to them in terms of verbal development and speech skills (for example, girls learn to speak and read earlier than boys).

Doesn't this make you think - no, not that men are smarter than women, what a sexist thing to say - but that our brains are a bit like a computing device with parallel processing in two autonomous centers - hemispheres?

Two personalities in one body

In medical practice, brain research is difficult for very obvious reasons - any intervention leads to irreversible consequences. Therefore, observations on the damaged brain are especially intensive during wars or on patients after serious traumas or with severe mental disorders. For example, hemispheric separation surgery is indicated in severe cases of epilepsy. One such patient was a unique case in the history of medicine.

At an early age, P. contracted epilepsy, as a result of which his left hemisphere became damaged and speech functions were partially taken over by the right hemisphere. After surgery to separate the hemispheres, doctors were able to ... communicate with both the left and right halves of his brain. The left hand, controlled by the right hemisphere, answered the medics' questions by making words out of individual letter cards. The right hand (after the researchers turned to the left hemisphere) wrote the answers or the patient said them aloud. Using a special testing technique, the patient was asked the question "What would you like to be?" and the left hemisphere gave the answer - "a constructor" and the right hemisphere - "a racer". Thus, two independent layers of consciousness were neighboring in P.'s psyche.

Researcher Mike Gazzaniga was shocked by the discovery he made. "One half of the brain," he later described, "was stating its own feelings and views, while the other half - the left half, which possesses speech - forgetting its dominant role for a while, watched its silent partner express its judgments..."

For each question addressed to the right side of the subject's brain using a special technique, the subject gave a detailed answer - for example, about his food preferences or favorite TV programs. When the left side of the brain was asked to answer the same question, the patient answered that he had not been asked about anything before, after which the right hand began to form the answer to the question from letters.

The most curious thing is that the left hemisphere carefully "observed" the actions that the patient performed at the command of the right hemisphere, and tried to justify them logically, to give a reasonable explanation for this or that answer!

For example, when with the help of a special device both hemispheres of the brain were simultaneously shown two pictures (rooster, winter) and asked to choose cards logically related to these images, the patient took two pictures with different hands: the right hand - a picture of a hen (the right hand is subordinate to the left hemisphere), and the left hand - a picture of skates (the right hemisphere was responsible for this choice). When asked what picture he was shown, the patient answered that he had seen a rooster, so he chose a chicken. And, after hesitating, he added quite confidently: "And a chicken is easy to clean with a skate." The left hemisphere immediately tried to justify the inference made, to logically connect the second picture with the choice it had made independently in response to the visual stimulus. But most importantly, the patient was not asked to give such an explanation. He was expressing his guess, trying to give a logical justification for his actions, to somehow harmonize the results of the actions of both hemispheres.

In psychology, there is a theory according to which a person tries to avoid disharmony between beliefs and actions by all means. For example, if a cautious and rational person is suddenly drawn into a crazy sexual relationship, he will constantly experience discomfort, feeling guilty, bad, standing on the edge of the abyss. To eliminate this dissonance, he will either break the connection (even if he was happy in this relationship), or will try to justify (to himself, not to others) such strange behavior - for example, that he is so trapped in the corporate framework that he needs a storm of feelings in his personal life. In other words, he will try to explain or justify this dissonance in any way he can.

Researcher Michael Gazzaniga believes (although many consider his theory speculative) that the external world and our own personality are constantly posing questions to our consciousness that simply cannot be answered, because none of us has complete knowledge or is omniscient. Therefore, the left hemisphere (which is responsible for speech and verbal expression of thoughts) tries to somehow more or less logically justify our actions and assure us of the wholeness of our personality, of the correctness of our ideas about ourselves.

Another brain researcher, Julian Jaynes, believes that the realization of the integrity of the individual emerged with the advent of writing and the first attempts at introspection. By the standards of historical science, this happened relatively recently - about three thousand years ago. And before that, according to Janes, man had a "bicameral" consciousness, when the hemispheres of the brain sometimes acted independently of each other. Moreover, if speech was generated by the right hemisphere, and perceived by the left, a person could well accept signals from the depths of the brain as the voice of the gods addressed to him (from the point of view of modern psychiatry - a perfect illustration of the symptoms of schizophrenia). In many ancient legends and legends we see one god (or spirit) ordering a person to do something, and another advising the exact opposite. It looks quite like an inner dialog of consciousness. But people, not capable of self-analysis, could not somehow classify the inner voices (demon whispering, guardian angel warning-saving) and simply obeyed them.

Jaynes was convinced that the human brain "is even more plastic and more capable of adapting itself to its environment than we have previously supposed...We may assume that the neural substrate of consciousness is sufficiently flexible to allow, on the basis of learning and culture, a transition from bicameral thinking to self-consciousness."

And here is how researcher Vernon Mountcastle defined the processes that take place in the brain and shape our unique personality: "...sensory stimuli flow into peripheral nerve endings, and their neural copies are sent to the brain, to the great gray mantle of its cortex," he wrote figuratively. - We use them to create dynamic and constantly updated neural maps of the external world, of our place and orientation in it, and of the events taking place in it. At the level of feeling and sensation, your images and mine are essentially the same, easily identified by verbal description or by a common reaction. But beyond that, each such image is linked to genetic information with accumulated individual experience that makes each of us unique and unrepeatable. On the basis of this integral experience, each of us constructs, at the highest level of our perceptual experience, our own very personal view from within."

Brain matter is superfast internet

The two hemispheres of the brain, connected by a bundle of nerve fibers, form the cerebral cortex, which is responsible for processing all sensory information entering the brain. In the occipital lobe of the cortex, visual signals are processed; in the parietal lobe - bodily sensations; in the temporal lobe - auditory signals, speech. There are areas responsible for processing olfactory signals, and areas where sensory signals of different types are integrated. Areas of the frontal lobes are responsible for motor functions (regulating movements), as well as for planning and anticipation. In the prefrontal cortex, thinking processes and integration of all incoming information take place, and all cortical areas continuously interact with each other and with subcortical structures (trunk, thalamus, etc.) and are connected into a single network (something like the Internet with constant access, super speed and unlimited traffic).

What is under the cortex and makes up the main substance of the brain? These are neurons (there are 5 to 20 billion of them in the human brain) - nerve cells responsible for processing information (gray matter is determined by the color of nerve cell bodies); nerve fibers that conduct information to different parts of the brain (the substance covering them, myelin, is white), and glial cells (there are 10 times more of them than neurons) that fill the space between neurons and form the "skeleton" of nerve tissue, as well as perform additional functions of metabolism and protection (e.g., recovery from infections).

From the body of the neuron, surrounded by a semipermeable membrane, there are nerve fibers - a long spur (axon), responsible for transmitting information to other cells, and short branching dendrites, through which the cell receives information from other neurons.

Information is transmitted from cell to cell by a nerve impulse that originates in dendrites (which receive information from other neurons) and travels from the cell body along an axon (the end of which can branch out to contact other neurons) at a speed of about 100 meters per second. The impulse is transmitted through a synapse (a narrow slit) by chemicals called neurotransmitters. At the end of the axon (a long fiber) are peculiar vesicles containing neurotransmitter. Having received an impulse, the axon releases the substance from the vesicle into the synapse gap, and this information is received by other neurons. Moreover, different neurotransmitters have only two differently polar actions - excitation and inhibition.

At rest, a neuron has a charge of 70 millivolts (the inside of the membrane is negatively charged relative to the outside). Dendrites (short fibers) conduct a neurotransmitter to the cell membrane, which helps enhance the conductive flow of potassium and sodium ions (excitation) or potassium and chloride ions (inhibition). Potassium and sodium ions, as they pass through the cell membrane, reduce the charge of its inner surface, and depolarization (excitation activity) occurs. If the negative charge of the inner surface of the membrane increases, hyperpolarization (inhibition) occurs. And then there is electricity! (So maybe it is for nothing to doubt the normality of people who tell how magnetic fields negatively affect them, or refuse to use cell phones).

Interestingly, the neuron, this microscopic, almost invisible particle, must integrate all the signals coming to it along the dendrites (and they can be directly opposite) and produce a single result.

You've probably already heard of the holy of holies, the most famous neurotransmitters in the brain. For example, serotonin, which causes a feeling of pleasure (it is said to increase if you eat a banana) or endorphin, which reduces pain and causes euphoria (ten minutes of dancing to your favorite music will increase the amount of serotonin dramatically). Other major neurotransmitters include norepinephrine, acetylcholine, glutamate, dopamine, enkephalins, and gamma-aminobutyric acid (many neurotransmitters have not yet been studied or named).

There are curious substances in nature that can affect the synthesis of neurotransmitters, their blockade, accumulation and release in vesicles at the ends of axons, or even mimic their action. For example, amphetamine accelerates the release of norepinephrine from vesicles, and LSD (diethylamidalizergic acid) can bind directly to serotonin receptors. Aminazine blocks dopamine receptors, morphine rapidly activates endorphin receptors. It is in this way - by strengthening or weakening the action of neurotransmitters - that psychotropic substances influence brain function, sensations, and even human behavior (Carlos Castaneda wrote about this in great detail, taking the psychotropic effect of peyote cactus as a basis).

How does it work?

And how does a normal, drug-free brain work? Let's take a simple example. You want to pick up a tube of lipstick from the dressing table. The light reflected from the tube is focused by the crystalline lens, arrives on the retina, then is transmitted by nerve cells to the visual center (thalamus); this is how light neurons are activated, which transmit the image to the part of the visual cortex (occipital lobe of the large hemispheres) from the right eye - located in the left, from the left - in the right hemisphere. Here begins a whole cacophony of impulses, some of which react to the shape of the tube, others to the shape and volume of the case, others analyze the boundary between the table and the lipstick. Then this information via axons goes to the analytical center, where all the data are brought together (we recognize the tube and the table). From there, the impulses go to the frontal lobes, where the planning of the movement begins (take!), then from the neurons through the nerve endings the signal goes to the muscles of the hand, to each finger (and the movement that has begun is controlled by visual receptors). And when we take the tube in our hands, the skin receptors of the fingers, reacting to the pressure, immediately report to the brain about how securely the fingers hold the lipstick and what degree of effort should be applied to hold it further. Well, in order to correctly put on lipstick or to come up with the theory of relativity, it requires such a quantity and quality of neurons that it is even scary to imagine it!

How can one not admire this amazing, perfect, complex and unique organ given to each of us? But we have only just begun to reveal to you the secrets of the brain....

CHAPTER II. CREATIVITY. HOW A GENIUS DIFFERS FROM AN ORDINARY MAN. WHY THE BRAIN WORKS DIFFERENTLY IN DIFFERENT PEOPLE

Is there no limit to human possibility?

Few people think about what the term "creativity" means. It seems quite ordinary to us, and we forget that only one creature on Earth can create - a human being. All others are deprived of this gift. Of course, everyone can remember the "painting" of monkeys, elephants and dolphins, but such "paintings" become creativity only in human perception.

The famous physicist, academician P. Kapitsa wrote that creativity can manifest itself in any field of human endeavor, where a person begins to act out of instruction. Animals or insects, no matter what beautiful dwellings they create for themselves, initially have an instruction in the form of instinct.

The essence of creativity is the discovery and creation of something new that has some value. New facts or laws in science, new devices in technology, new artistic images or artistic forms in art.....

Social psychologist, Englishman Graham Wallace in 1926 identified four stages of creative thinking:

- preparation - formulation of the problem, attempts to solve it;
- Incubation is a temporary distraction from the task at hand;
- insight - the emergence of an intuitive solution;
- verification - testing and/or implementation of the solution.

Each of us is familiar with these stages, no matter in what area we were creating: inventing something at work, painting a picture, composing a verse greeting for a loved one, figuring out how to fry potatoes more deliciously or how to paste wallpaper better. Any of us created. But why the creativity of some is limited to the kitchen and greeting cards, and others create masterpieces that remain in the centuries? Where in our brains is the button that a person can become a genius by pressing?

Is genius the mass of the brain or the number of twists?

For many centuries people have been trying to unravel the mystery of genius. Not only do we not know where it comes from, but often we cannot even articulate what it is. According to the definition of the English poet Coleridge, who was considered a genius by many, genius (at least on the outside) is the ability to grow. But what happens on the inside...? This question, it seems, could not be answered by Coleridge himself.

Dictionaries define "genius" as the highest degree of manifestation of human creative forces. Believers believe that genius is the birth of the Spirit of Truth in a person, but they cannot explain why God has given genius to some people and "grayness" or even stupidity to others.

Unlike religion, science has always tried to find out where genius comes from. For a long time it was believed that the secret lies in the number of twists in the brain of geniuses or even in the increased weight of the brain.

But, as practice shows, weight has nothing to do with it. In Russia, the largest brain was in the most talented Russian writer Ivan Sergeyevich Turgenev (1818-1883), and it weighed 2012 grams (the average weight of the brain of an adult man - 1375 grams, and an adult woman - 1275 grams). Only one larger brain is known, which weighed

2850 grams. It belonged to a patient in a European psychiatric hospital, an epileptic idiot.

Not all geniuses had "heavy brains". For example, the brain of the French writer Anatole France weighed only 1017 grams, which did not prevent him from forever entering the history of world literature.

Let's compare the brain weights of famous people and once again make sure that it is impossible to deduce any regularity here:

Sergei Yesenin, Russian poet - 1920;

George Byron, English poet - 1807;

Otto Bismarck, German politician - 1800;

Ludwig van Beethoven, German composer - 1750;

Vladimir Mayakovsky, Soviet poet - 1700;

William Thackeray, English writer - 1658;

Ivan Pavlov, Russian scientist - 1653;

Immanuel Kant, German philosopher - 1600;

Dmitri Mendeleev, Russian scientist - 1580;

Lev Trotsky, revolutionary - 1568;

Konstantin Stanislavsky, artist - 1505;

Andrei Sakharov, scientist and politician - 1440;

Robert Kennedy, American politician - 1432;

Marilyn Monroe, American movie star - 1422;

Franz Schubert, Austrian composer - 1420;

Maxim Gorky, Soviet writer - 1420;

Dante Alighieri, Italian poet - 1420;

Alexei Tolstoy, Soviet writer - 1400;

Konstantin Tsiolkovsky, Russian scientist - 1372;

Vladimir Lenin, revolutionary - 1340;

Albert Einstein, scientist - 1230.

It is known that the size of the brain is partly proportional to the growth of the body, and therefore many people, speaking of Turgenev's brain, remember not only his huge for that time growth, as much as 192 centimeters, but also the total mass. Vasily Rozanov, for example, wrote: "Physiologically disgusting Turgenev consisted in what is not visible in the portrait, as we know - rather striking beauty, harmony and grace: his large head sits not only on a larger, but - immeasurably larger body, which is exactly all swollen like a dead drowned man. His body, his figure is so huge, as if it were some kind of furniture, and not a living, real man..." But even this theory can be accepted only with great reservations: the second largest brain of Russian literature belonged to Esenin, who, as we know, was very modest height: only 168 centimeters.

By the way, the size of the brain also depends on the nationality. The most "brainy" are Eskimos, their average brain weight is 1558 grams. The other nations are distributed as follows:

Buryats - 1524,

Hindus - 1514,

Polynesians - 1475,

French - 1473,

the English - 1456,

Chinese - 1430,

Russians and Ukrainians - 1377,

Japanese - 1374,

Europeans (average for all European countries) - 1361,

Swiss - 1327,

Negroes (African continent) - 1316,

Germans - 1291.

As we can see, it is impossible to deduce any regularities here. So science has long ago abandoned the idea that brain weight affects talent. Now the theories that genius depends on the quantity and quality of crinkles, as well as on the total number of neurons, are more in vogue. But these are, after all, too general. There are, however, other theories.

We know that our brains are far from working at full capacity. At least that's what scientists suggest. Some believe that the human brain uses 10% of its resources, some that 20%, the most courageous suggest that the brain puts at our daily disposal as much as 30% of its power. Sometimes they draw an analogy between the power of thought and the power of the physical and recall events where people have shown tremendous power in extreme situations. For example, in Italy, a car hit a girl, and her mother, an ordinary woman who had never played sports, simply turned the car over and pulled her child out from under the wheels. Many assume that if the same extreme way to launch all 100% of the brain power, then just and you can become a genius. The main thing is to find that magic button. But scientists, however, do not share such optimism. Rather, on the contrary, they believe that for genius it is necessary not to "start" the reserve parts, but to "turn off" the working ones.

Idiots are geniuses

Psychologists and psychiatrists have long known so-called "idiot geniuses", mentally retarded people who show tremendous ability in any field of knowledge. Suffice it to recall the Hollywood movie "Rainman" (in Russian translation - "Rain Man"), where Dustin Hoffman played a patient with autism. His character cannot navigate the world around him, but he can easily count several hundred fallen toothpicks in a fraction of a second, easily count cards in a casino, and so on.

Real, non-movie autistic people surprise others much less often, but it does happen. Many of them have superpowers in math, music and other areas of science and culture. In scientific circles is widely known autistic person who invented 24 languages and perfectly communicates in them. As linguists say, these are high-quality, deeply developed languages, and any of them can be used as a means of international communication. Practice shows that the thorough development of artificial languages takes decades of labor of many people, and this man single-handedly invented two and more than a dozen languages ... But the IQ test, which reveals intellectual ability, such an inventor would fail with a bang. Like other autistic people who can, after looking at a house, immediately draw its detailed architectural design, learn the city telephone directory in a day, or simply name with accuracy to a few millimeters all the dimensions of, for example, a table or a cabinet.

Scientists see no mystery in the presence of such abilities in autistic people. A certain part of their brain is fully engaged in its work, and it is not confused or distracted by thoughts from other parts of the brain. Many experts believe that anyone can replicate the achievements of autistic people.

Allan Snyder and John Mitchell from the Center for the Study of the Mind at the Australian National University have conducted many studies of human superpowers and devoted several works to their disclosure. According to Australian scientists, we, i.e. ordinary people, automatically try to **comprehend** facts and observations, while an autistic person (or in this case we should rather speak about a genius) is not distracted by this, dealing only with bare facts and not moving to generalizations and conclusions.

Snyder and Mitchell believe that the responsibility for conclusions and generalizations in the brains of such geniuses is borne by lower, evolutionarily older parts of the brain. However, they also function in ordinary people, but they are "blocked" by more highly developed parts.

Massive research has suggested what is the essence of the difference between the perception of reality by an autistic person and a normal person. It is known that the image focused by the lens of the eye reaches the retina and is then sent to the brain. Between the picture on the retina and its comprehension passes only about a quarter of a second, but during this time a tremendous amount of work is done: specialized areas of the brain identify each of its own: color, size, distance to the object, the speed of its movement, and so on. Then all this is transmitted to the higher parts of the brain, and along the way all kinds of technical "garbage" is sifted out. Otherwise, we would not be able to breathe from the abundance of details and unnecessary data in this particular case. Therefore, only what is necessary to make a decision about what we have seen is supplied to the higher parts of the brain.

In autistic people, all this data comes without any processing, with a lot of tiny details, and therefore such a person can easily remember, for example, all the names from the phone book: all this data has not been sifted out by his internal filters and lies on the "surface" of the memory. Naturally, the memory of an ordinary person also retains everything once seen, but it is not easy to get to this information: our brain, afraid of overloading, puts the most powerful barriers.

Although there are known cases when with the help of hypnosis the smallest nuances of events that took place several years ago were retrieved from human memory. For example, there is a case when a witness under hypnosis remembered the license plate number of a car he had just passed by on the street three years ago. The person could not even suppose that he paid attention to this car, but, however, even its license plate number remained in his memory, and the criminal offense was solved. This is almost a classic example, but modern investigators also do not forget about hypnosis. For example, it was used in the investigation of the murder of the director of one of the Ryazan joint-stock companies. When the investigation had at least some data, it was three years after the tragedy, and the wife of the murdered man, the only witness, has thoroughly forgotten that terrible day for her. But under hypnosis, she remembered how the killer was dressed, and how he looked, and many other useful details. The crime was solved.

But don't think that memories stored somewhere in the depths of the brain have no impact on our lives. Psychotherapists say that nothing goes away. That is why psychoanalysts and try to "dig" to our childhood, claiming that the roots of all problems lie precisely in childhood fears. For example, a good specialist can take an adult back to his childhood and make him remember the day he was bitten by a dog. In this case, the psychoanalyst is able to "change the scenario" a little, and instead of experiencing the shock of the bite, the child will simply pet the dog. After such a "trip," a patient is known to have recovered from a stutter that he had been unable to get rid of for several decades.

But besides the "garbage" of memories, our brain also blocks a lot of our abilities, seeing no practical sense in them (and just trying not to overstretch). For example, if you now ask yourself a question about what day of the week Pushkin was born (it was June 6, 1799), then, understandably, to get the answer, you will need a calculator and a few minutes of calculations (or a perpetual calendar). But in fact, your brain has already done the instantaneous calculations and answered the question. It just didn't put the answer "on the display" because it didn't see the practical need for it.

But in a stressful situation, when your brain realizes that your life depends on the answer, there is a good chance that you will answer the question correctly without a calculator. That is, this barrier, which is absent in autistic people who instantly perform the most complex mathematical calculations, can be removed. But how do you do it without going to the extreme?...?

Note that world science knows many discoveries made in dreams. Among Russians, by far the most famous dream is Mendeleev's table. In such cases, everything is very simple: the natural obstacles of the brain did not allow Mendeleev to discard the usual logic and the categories with which he was accustomed to think, looking at the problem, the solution to which he was struggling, from a different angle. In the dream, these logical "slingshots" disappeared. But still, this example, although clear, is not quite legitimate: it is much easier to remove the "slingshots" of logic set up by an individual himself than those that our brain has set up by itself.

Pushkin, by the way, was born on a Thursday.

Did you ever have a thought about this day? If your answer was correct, it could have come from two sources: your brain either calculated this day, or simply remembered information you had once heard.

But back to autistic people. Many scientists believe that the root of their superpowers lies not in the absence of barriers, but in the fact that some parts of their brain perform functions that are not typical for them. According to this theory, all the reserves of an autistic person's brain are focused on a single task, such as mathematical calculations, and the part of the brain responsible for this function is even physically stronger than the others.

In most people, according to scientists who adhere to this theory, such a one-sided development does not occur because the brain from an early age is tuned to generalization and conclusions, ie focuses on the big picture, rather than fixing individual details. In autistic people, however, internal connections in the brain are weakened, and therefore instead of a general impression "on the scoreboard" gets a lot of nuances that pass by an ordinary person.

Deprived of the need to sift out the unnecessary, the inner brain zones begin to "boil in their own juice" and develop, making a person a genius, for example, in verbal counting, in visual memory, and so on. But, alas, an idiot in everything else.

Our science, unfortunately, is not yet so developed that it is possible to clearly accept one version and reject the second. There are quite reliable facts that speak both in favor of the first theory and the second. For example, supporters of the second version like to recall the results of the study of Einstein's brain. Those areas of his brain, which were "responsible" for mathematical ability, were not only increased in size, but also did not cross the gyrus, as in ordinary people. It is believed that the gyrus "limit" different areas of the brain, and, as scientists suggest, the "mathematical segment" in Einstein's brain has swept under itself neighboring areas that were not at all for mathematics. A significant drawback of this example and others like it is that the brain was examined after the scientist's death, and therefore we cannot say whether the "mathematical segment" was such from birth, or it acquired its form after decades of calculations.

In favor of the first version, which states that every person is genius from birth, says, for example, the fact that an eight-month-old infant unconsciously performs fantastically complex calculations that allow him to understand where in the stream of speech ends one word and begins another. And soon the infant begins to speak. But it takes at least months of hard work for an adult to learn a foreign language.

It is no secret that children learn much better than adults, but these superpowers to learn new things disappear with age. Studies have shown that newborn children and infants have active parts of the brain that "sleep" in adults. It is thanks to their activity in the child develops "automatic" behavior. For example, a four-year-old child would hypothetically need only a few hours to learn to drive a car, while an adult would take several days to several months, depending on age.

The higher part of the brain, responsible for intelligent behavior, "switches on" only a few months after birth and gradually gains strength, reaching its maximum at the age of about one and a half years, when children begin to speak. Autistic patients either do not have this switch at all, or it is only partially performed, and therefore the manifestations of the amazing abilities of the "infant" brain amaze us in adults.

Usually, autistic people have a poorly functioning left hemisphere, which is responsible for logic and rational behavior, while the right hemisphere, which governs intuition and emotions, is overdeveloped. In known cases of autistic people being cured, their left hemisphere eventually took over and all superpowers were completely lost.

Lombroso's research

Let's get away from the extremes and talk about those who, being quite oriented in our world, are still considered geniuses, skillfully combining unique abilities with normal brain development. (However, the famous Italian psychiatrist Cesare Lombroso considered genius to be a form of insanity).

It is known that the main "enemies" of creativity are fear of failure and high self-criticism. It is these features do not allow many people with extraordinary abilities to go on the waves of creativity. And that is why since antiquity, many philosophers have drawn a parallel between genius and damage to reason. The founder of philosophy Aristotle noticed that, according to his observations, under the influence of tides of blood to the head, many people "are made poets, prophets or soothsayers," and cited as an example that "Marcus of Syracuse wrote quite good poetry while he was a maniac, but, having recovered, completely lost this ability". In his other work, Aristotle again returns to the topic of interest, saying that many famous politicians, poets and artists "were partly melancholic and insane, partly misanthropic, like Bellerophontes. Even at the present day we see the same in Socrates, Empedocles, Plato, and others,-and most strongly in poets. People with cold, abundant blood are timid and limited, while people with hot blood are agile, witty and talkative."

Plato also supported his colleague, noting that "delirium is not a disease at all, but, on the contrary, the greatest of the blessings given to us by the gods; under the influence of delirium Delphic and Dodonian prophetesses rendered thousands of services to the citizens of Greece, while in the usual state they were of little use or completely useless. Many times it happened that when the gods sent peoples epidemics, one of the mortals fell into a sacred delirium and, becoming under the influence of his prophet, indicated the cure for these diseases. A special kind of delirium, excited by the muse, causes in a simple and pure soul of man's ability to express in a beautiful poetic form of heroes' deeds, which contributes to the enlightenment of future generations". He was echoed by Democritus, who directly stated that can not consider a true poet of a person who is in his right mind.

The impracticality of geniuses was also noted later. The French religious thinker, mathematician and physicist Blaise Pascal (1623-1662), considered one of the greatest thinkers of his century, liked to say that the greatest genius borders on complete madness, and, alas, proved this postulate with his biography.

And the French literary man, lexicographer and local historian Gabriel Hécart (1755-1838) published in 1823 a book entitled: "Stultitiana, or Brief Bibliography of the Lunatics in Valenciennes, compiled by a lunatic." Eckard's poems are still considered gems of poetry and, at the same time, the work of a man in a borderline mental state.

Lombroso gives many examples of geniuses behaving in such a way that, if they were ordinary people, they would at least be treated for a long time.

The French naturalist Leclerc de Buffon, immersed in his reflections, once climbed to the bell tower and descended from there on a rope, not realizing what he was doing, and then did not remember it.

The great French physicist André-Marie Ampère could not speak except by walking and moving all his members.

Many geniuses had very strange facial expressions that could be mistaken for grimaces. Napoleon suffered from a constant twitching of his right shoulder and lips, and Peter the Great, when he was in an excited state, suffered from violent spasms of the facial muscles.

Lombroso also quite rightly remarks that the pathological state of the brain often acquires external manifestations: baldness or early graying, thin body, poor muscular and sexual activity. In addition, all these physical signs, peculiar to the insane, are often found in great thinkers. Segur wrote about the French writer, philosopher and historian Voltaire: "Strong thinness proves how much he works, bent and weary body of his serves only light, almost transparent shell, through which as if you see the soul and genius of this man. You yourself, probably, can list a lot of geniuses whose physical development can hardly be attributed to the norm. The first to come to mind is the short and stooped academician Sakharov, the inventor of the hydrogen bomb, who was often called "dandelion" because of his gray hair and stooped physique.

We can recall another sign of geniuses: strange relationships with the opposite sex. And if Michelangelo, for example, said that his art replaced his wife, then Heine, Byron, Goethe, Napoleon, and Newton, as Lombroso notes, "proved something even worse by their actions. We can remember Pushkin's "donjuan list" and similar lists of many other Russian geniuses. A psychologist or psychiatrist often interprets what an ordinary person considers a dalliance or luck as an anamnesis, and he is right.

We can also recall Dostoevsky's passion for roulette (who, by the way, is considered by some, along with Nabokov, to be a latent pedophile); Lermontov's pathological acrimony; Tolstoy's spiritual quest; Mayakovsky's numerous oddities (fearing contagion, he opened doors with his feet, and if it was impossible to do so, stood and waited for someone to come out); Yesenin's alcoholism and madness (about his oddities in his last days wrote a lot close friend of the poet, also a poet, Anatoly Mariengof); Vysotsky's drug addiction.... In fact, digging into any genius, we will find some pathology in his biography. Many people rightly attribute the appearance of these oddities, as well as talent, to the consequences of "wrong" brain development.

It is even more interesting to observe how geniuses became initially quite ordinary people. And the impetus for this was brain trauma. Giambattista Vico, the greatest Italian philosopher of the Age of Enlightenment, the creator of the modern philosophy of history, who laid the foundations of cultural anthropology and ethnology, fell from a high ladder as a child and shattered his right parietal bone. However, we have no evidence that he was not originally a genius, but he himself considered this fall to be the reason for his talent. There are more convincing examples: a weak singer Gratry, who moved in provincial theaters, after a strong blow log in the head became a star of the first magnitude. It is clear that the bruise changed not the vocal cords, which remained intact, but only the ability to manage them. Hence, by the way, we can draw another conclusion: having found a genius button, you can not only learn to count quickly or develop memory, but also begin to better manage the functions of your body.

There are more surprising cases: French scientist, historian, Benedictine, the "father" of the science of paleography and diplomacy Jean Mabillon (1632-1707) in his youth was completely demented, and having received a severe head injury, not only wised up, but also became a light of reason.

There is a similar case in Russian history: Metropolitan Macarius (Bulgakov) of Moscow (1816-1882), a famous church historian and theologian (a famous Orthodox award is named in his honor, and textbooks written by him are still taught today), was a very sickly, stupid child and could not study at all. Other children, seeing his weakness and underdevelopment, teased and cruelly persecuted him, and once, when the boys threw stones at the future metropolitan, one of the cobblestones cut "the cranial shell on his head, causing severe bleeding". Surprisingly, but after that the boy not only discovered outstanding mental gifts, but also completely improved his health.

Inspired by the universe

It is interesting that the creative process itself in many geniuses occurs in a state that can hardly be called normal. As Jürgen Meyer correctly observed, a simply talented person acts in a strictly deliberate manner; he knows how and why he came to his conclusions, whereas a genius is completely unaware of this: his creative activity is, most often, unconscious.

The same Napoleon admitted that the outcome of many battles won by him, often depended on a single moment, from one thought, which, at the onset of a favorable moment "flashed like a spark", and the result of such insight became a victory.

Geniuses very often say that their works are "dictated from above" - allegedly they themselves cannot understand how their creativity is born; but a simply talented person will always be able to describe how this or that creation was born - the whole chain of thoughts, plans, accomplishments.

When Haydn wrote the famous oratorio "The Creation of the World" and he lost inspiration, he went with rosary beads in his hands to the prayer room, where he called on the Virgin Mary - after that the inspiration returned again.

Milli, the famous Italian poetess, said that while composing her poems it was as if she were experiencing an epileptic seizure: worrying, shouting, singing, running back and forth, completely unaware of her actions. Even Goethe wrote that the poet needs "a certain brain state", and admitted that he himself composed many of his works, being as if in a state of somnambulism.

Count Vittorio Alfieri, Italian poet and playwright-classicist, "the father of Italian tragedy", called himself a barometer - so much so that his creative ability varied depending on the season: starting in September, he could not resist the feeling that possessed him and had to create six comedies. And on the manuscript of one of his sonnets he wrote: "Accidental. I did not want to write it."

"All works of genius," wrote Voltaire to Diderot, "were created instinctively. The philosophers of the whole world together could not have written Armida Kino or the fable "The Pestilence of Beasts," which La Fontaine dictated, not even knowing well what would come out of it. Corneille wrote the tragedy "Horace" as instinctively as a bird builds a nest".

In fact, we see that the creative process of genius can be compared to a certain mental disorder, which the person himself is unable to control. For example, with kleptomania, when a well-to-do person steals things he does not need, knowing that he can pay for this weakness with his position and fortune. After all, even genius often came out to its owners sideways.

The only thing left for mankind is to try to separate overwork of the brain from the "bad consequences" that accompany it. But is it possible? Probably not. So genius will remain forever an unattainable peak of human brain work. But we believe that talent and ability can be developed. Moreover, this has been proven by numerous scientific experiments.

"The Farm" of Talent "

More than twenty years ago, a global experiment to breed geniuses began in the United States. In 1980, genetic scientist Robert Graham, with the support of a number of foundations and individuals, began collecting sperm from the most outstanding people on the planet. The main part of the bank was made up of donors - Nobel Prize winners. Their names are unknown, but the winner of the prize, the legendary mathematician John Forbes Nash, praised in the movie "Mind Games", himself admitted that he was involved in this project. Jonas Salk, one of the creators of the polio vaccine, was also likely involved.

But, as it turned out, women do not want to give birth to geniuses at all, they are more interested in the child being beautiful and healthy. This became the main difficulty of the project. Nevertheless, until 1999, more than 200 children were born within the framework of the project. And, it turned out... that all of them turned out to have the most ordinary data, in no way resembling their fathers. The only one who could be called a genius - Doron Blake, whose IQ was 180, and who at the age of two already knew how to use a computer and at the age of five read

Hamlet - after a few years of college, abandoned science and became interested only in the occult and drugs.

The experiment, although it was a failure, still gave certain results: we once again became convinced that the matter is not so much in genes as in the structure of the brain, the dependence of which on genes, alas, cannot be conducted directly.

How to develop creative thinking and become a creative person

Alexander Khayem, a Harvard undergraduate and corporate consultant to General Motors employees, identifies thirty of the most significant factors that directly influence people's use of their creative potential. Among the most important he names knowledge of basic methods of creative thinking (generating ideas by word association, brainstorming) and constant study of specialized literature.

An important factor is a sense of control over the situation, which is determined by a positive attitude (work on your mood: depression is the first enemy of creativity) and the memory of past achievements (which should be reminded more often).

The ability to openly accept alternative viewpoints is one of the most important components of creative thinking. Take every opportunity to communicate with strangers, especially those you dislike. Take an interest in their opinion on important issues to develop tolerance for other people's positions, the ability to follow the thought process of another person (of course, this all applies to people close to you in intellectual ability; do not waste your time on dullards). And if you are lucky and among your acquaintances there are people who have achieved success, thanks to their personal creative abilities, try to communicate with them as much as possible. This can seriously increase your own creativity (by the way, a successful entrepreneur is a person who is at least creatively talented, although he may not even suspect it). Adopt their behavioral patterns, observe how they work.

The more you learn about new people, ideas, books, movies, places, the more you experience the world in all its diversity, the more material you have for creative thinking. And don't forget that being too serious and heavy-handed in your approach to fresh ideas can kill creativity - every now and then try to treat new tasks as a fun game that you can get the most out of. Look for people who have a lighter approach to life than you do, and when you have a period of creative stupor - try to socialize with them as much as possible. But at the same time, don't forget to persevere in achieving your goals. If you're used to abandoning things halfway through, set yourself specific goals that are time-bound.

Practice creativity in the most ordinary things - think of a new route to work, do a new makeup or hairstyle every day (men probably shouldn't practice this option), make a new coffee recipe every morning. By developing a creative approach to everything in the world, you will soon be able to easily generate innovative ideas at work.

Engage in self-reflection - analyze how you create, what you usually do, what objects, skills or strategies you use. Once you are aware of all these factors, change your routine dramatically: try creating earlier or later than usual, or in unusual environments - on the way to work, in the bathroom, or late at night. Try putting on music or working with headphones that isolate you from outside noise. See if your productivity improves in a different environment. Look for the best times and the most suitable conditions to express your creativity.

Do not be afraid of creative failures - for a boring or banal idea you will not be shot as a traitor to the motherland or sent to the Gulag. Remember, you are not at the operating table and can easily afford to make a mistake - there is a certain amount of happiness in the very essence of the creative process. You won't destroy the project - you'll just spend a little more time on it. So be brave.

A simple psychological trick helps to develop creative thinking - imagine yourself as a person who admires you for your talent (a famous artist, director, actress). In a situation where you need to be creative, imagine that you are reincarnated in one of these personalities. Ask yourself a question - how would he (or she) act, what would he (or she) do, how would he (she) approach the problem. Imitate creative people, and you will soon become creative yourself.

CHAPTER III. STIMULANTS OR KILLERS? CAFFEINE, ALCOHOL, AND DRUGS-HOW THEY AFFECT THE BRAIN. THE SECRETS OF SIBERIAN SHAMANS

Man has long been picking up the key to his brain, hoping with a slight gesture, a half-turn to adjust this complex mechanism so that it generates on a predetermined topic and that these thoughts struck with their depth and power.

Prayer and meditation

Prayer was the first attempt of mankind to make the brain think in the right direction. Even in our days, studies have been conducted which have shown that in students of the Theological Academy and seminary during prayer there was a gradual weakening of the rhythm of the brain's biocurrents up to the dominance of only slow delta rhythms. In this case, the actually awake person was in a state of so-called slow-wave sleep. Such a state of the brain occurs during wakefulness only in infants, and that only up to two months: Electroencephalogram showed in the most diligent praying almost complete shutdown of the cerebral cortex, although they were fully conscious! Interestingly, when a person stopped praying, the rhythm of electrical impulses of his cerebral cortex immediately returned to the biological norm. Such a state, according to experts, helps to destroy pathological connections both in the brain and in the body, to produce, in computer language, a reboot. In the previous chapter we talked about the fact that when Haydn lost his inspiration, he prayed and could write music again. The great composer simply... overloaded his brain, clearing the "RAM" of unnecessary patterns.

Eastern meditation is also very close to Christian prayer. There are many kinds of meditation, and not all of them have been fully explored. For now we can only say that the state attained in some types of meditation is something between wakefulness and drowsiness.

After a meditation session there is a decrease in physical and mental tension and an increase in general mental activity. One of the tasks of meditation and one of its advantages is a kind of "emptiness" of the mind - a state that can hardly be described in words and is very important for clearing the mind of "templates". Today medics are trying to treat various psychological fears and addictions with the help of meditation. Taking into account that no ideal panacea has been invented yet, we can say that this method is not worse than others, especially since sometimes it has the desired effect.

A little digression: a hormone called melatonin plays a significant role in our lives. It is produced by the epiphysis, a pineal body located deep beneath the cerebral hemispheres. We will not go into scientific terminology and details, but just say that melatonin performs a number of very important functions in our body:

- regulates the endocrine system;
- slows down the aging process;
- enhances the efficiency of the immune system;
- has antioxidant properties;
- normalizes and restores full sleep;
- eliminates and mitigates the effects of stress.

It also, more interestingly in the context of this book, regulates brain function and improves performance. Melatonin is also known as the "happy hormone".

You have probably already realized that if the body lacks this hormone, a person becomes more susceptible to stress, his sleep is disturbed, and many processes in the body fail. But, as the French researchers found out, you can make the body produce melatonin. To do this, it is necessary to pray. Studies that were conducted by French scientists were based on reading the so-called Jesus Prayer, which was recited exactly as commanded by the first teachers of Christianity: on the inhale should have said "Lord Jesus Christ, Son of God ...", on the exhale "...have mercy on me, a sinner!". By the way, this very prayer was also recited by Russian seminarians at the request of Russian scholars. The point here, as researchers claim, is not in the words spoken, but in the state into which the brain falls when reciting this brief formula according to certain rules.

The important thing is that the prayer is not read mentally, but aloud. And this is not by chance: modern research has shown that chewing stimulates the hippocampus, the area of the brain responsible for the assimilation and storage of information. It is clear that breathing is regulated for a reason: a steady supply of oxygen to the brain has a positive effect on thought processes.

Notice how interesting it is: we write that "French scientists have made a discovery," but in fact such an effect of prayer was known to our ancestors many centuries ago!

Mantras similar in effect to the Jesus Prayer can be found in Buddhism, Hinduism, and other religions. They are the first way invented by man to influence the brain. It was worth to concentrate a little, to say the cherished words, and the psyche immediately calmed down, thought processes were activated, and efficiency increased. But a person always wanted more. Well, one would think that he began to think a little more productively, but if the brain power were to increase several times at once!

And mankind turned to chemistry. In those days, they did not know about chemistry yet, but various plants could be collected right under their feet. Anyone has heard of ginseng, eleutherococcus or lemongrass - the most popular brain stimulants.

And we also try to stimulate ourselves every day with very simple methods: with tea and coffee, and some also with smoking.

Green tea is known to have the best effect on the whole body and the brain. Recently, mate tea has become popular, as it contains a lot of vitamins and amino acids and has the ability to concentrate attention and improve short-term memory. But many people do not recommend drinking coffee: despite the strong stimulation, the effect of this drink is short-lived, and the boost is quickly replaced by a long decline.

The main component of these drinks that affects us is caffeine.

Caffeine

In medicine, caffeine is known as trimethylxanthine. It is a white crystalline powder with a very bitter taste. It is used as a cardiac stimulant and diuretic. And, of course, to induce increased activity. Caffeine is essentially a weak drug. It affects the brain in the same way as its "older" colleagues, and it is very easy to get hooked on coffee. Although, of course, both attachment and withdrawal in case of refusal will be only a pale shadow of similar consequences when taking cocaine or heroin.

There are 90 to 200 milligrams of caffeine in a cup of ground coffee, 30 to 70 milligrams in a cup of tea, and 30 to 45 milligrams per cup in various colas (Pepsi, Coke, etc.).

Under the influence of caffeine, the brain releases so-called adenosine, which, by attaching to receptors, causes drowsiness and slows down the activity of nerve cells. It also causes the brain's blood vessels to dilate so that the brain can absorb more oxygen during sleep. But the thing is that receptors confuse adenosine and caffeine, and once caffeine has joined the receptor, adenosine has no place there, and the brain, deprived of a permanent natural brake, begins to "accelerate". When the blood vessels stop dilating, the pituitary gland, seeing that there is something strange going on in the brain, decides that the reason for this is some circumstances from which it is better to be safe, and a hormone goes into the bloodstream, forcing the adrenal glands to generate adrenaline, which puts the body in a state of "combat readiness". Pupils dilate, allowing better vision, breathing becomes more rapid, providing excess oxygen, heartbeat increases to distill the oxygen received into the muscles faster. The liver releases sugar into the bloodstream to strengthen the muscles. Something similar happens to our brain: its blood vessels dilate, oxygen saturation increases, and the brain is ready for quick reactions and multiple tasks.

But soon the body "realizes" that nothing terrible is happening, adrenaline release stops, there comes a slump and fatigue, although, as it seems, we have not yet had time to do anything.

Approximately the same reaction is accompanied by the use of tea, but only with the exception that tea has less caffeine, and the reaction itself, as well as the "withdrawal" after it, is weaker.

These stimulants actually give our brains a slight advantage, but alas, for a very short time. The price for this boost is a loss of energy, when perhaps the work is not yet done and the tasks have not yet been completed.

A similar pattern occurs when the brain is stimulated by tobacco. Once nicotine reaches the brain, it acts on certain groups of neurons, triggering acetylcholine receptors in the absence of acetylcholine itself. The brain begins to release a number of substances, particularly noradrenaline and serotonin. Noradrenaline raises blood pressure, increases heart and respiratory rate, increases concentration and performance. Serotonin gives emotional stability, which helps to concentrate without being distracted by extraneous thoughts. Soviet scientist A. E. Shcherbakov conducted a large-scale study and proved that small doses of nicotine increase the excitability of the cerebral cortex, but for a very short time, and then depress and deplete the activity of nerve cells.

Nootropics

Recently, a number of drugs have appeared, so-called nootropics, or nootropics, or neurometabolic stimulants, the effect of which is much longer, and the harmful effects of their use are much less. This is, at least, what medics say.

The term "nootropic" itself is made up of the Greek νσυζ - mind and τρσπή - change. It appeared in 1972 to describe the effects of piracetam. Later on, similar effects were noticed with other substances.

Piracetam, now mostly known by its commercial name "nootropil", was synthesized by Belgian pharmacologists in 1963. Like psychostimulants, it enhances mental performance but does not have the inherent side effects of other psychostimulants.

Today, more than ten original nootropic drugs of the pyrrolidine series have been synthesized, either in the phase of clinical trials or already registered in a number of countries. These are oxiracetam, aniracetam, rolziracetam, etiracetam, detiracetam, pramiracetam, nefiracetam, dupiracetam, cebracetam, and izacetam. These nootropic drugs have the generic name racetams.

The brain has not yet been fully studied, and it is impossible to state anything with complete certainty, but it is assumed that although nootropics do not have a single mechanism of action, their effects may be caused, in particular, by a direct effect on neurons, improving cerebral blood flow and microcirculation in the brain.

Neurometabolic stimulants are thought to have both a nootropic effect (i.e., they elevate judgment, improve attention, thinking, and speech) and a mnemotropic effect (i.e., they affect memory and increase the ability to learn new material).

Nootropics also have negative properties: their stimulating effect on mental activity may be accompanied by speech and motor agitation, restlessness and sleep disorders. Sometimes addiction develops, but coffee and cigarette drinkers are hardly surprised by this.

Recently, there has been a lot of discussion in society, including in medical circles, about the possible use of nootropics by healthy people. It is one thing when these drugs are used to improve memory in stroke victims or people with mental retardation. Another is when a person wants to simply "speed up" his brain, not realizing the possible long-term consequences of using nootropics. These drugs have already been aptly nicknamed "Viagra for the brain," and, just as in sex, there is an ethical problem. Leon Kass, head of the U.S. President's Council on Bioethics and a moral philosopher, believes that "in those areas of human life where people have hitherto achieved success solely through discipline and dedication, the gains made by pills, genetic engineering or technical transplants smack of fraud or cheapness.

Many scientists agree with him. Jerome Yesavage of Stanford University and Peter Whitehouse of Case Western Reserve University conducted an entire study on the effects of donepezil, one of the many drugs for halting progressive memory loss in Alzheimer's disease, on the professional skills of pilots. Participants in the experiment were divided into two groups. One group took the drug and the other group took only a placebo, i.e., pacifiers. Both groups had to perform on simulators the most complex maneuvers and simultaneously respond to periodically arising abnormal situations. After a month, those who took the drug regularly began to show significantly better results.

Yesavage suggests that when it is definitively proven that intellectually competent people can improve their mental abilities with chemicals, serious legal, administrative, and ethical issues will arise. Indeed, it will be more convenient for employers to hire those who cheer up with pills rather than coffee. The race of medical drugs will begin, and it is hard to predict how it will end. But it's safe to say that the world is once again on the verge of a major redistribution, similar to the one that brought women who had previously stayed at home into the workplace. But the moral aspect is not so simple.....

But it is possible that governments, fearing the consequences, will ban people from chemically "boosting" their brains. After all, a similar story has already happened with the use of cocaine.

Relax!

Thousands of years ago the Indians discovered an amazing method of stimulation - they put some coca leaves, a lump of lime in their cheeks - and it quickly quenched their hunger, made them less sensitive to the cold of the high mountains, and, of course, lifted their spirits. It is clear that with such daily stimulation the average life span of the Indians was about thirty-five years, and even at that age they already looked like profound old men.

Realizing this dependence, the Indians themselves began to impose various restrictions on the use of coca leaves. In the Inca Empire, for example, coca became the privilege of the "children of the sun", i.e. the ruling class, and in some ethnic groups its use acquired a sacred meaning and became the prerogative of sorcerers.

The arriving Spaniards quickly came to appreciate the Indian "fun" and began to indulge in the leaves as well. They also gave coca to slaves to increase their productivity. For example, in the silver mines in Potosí (Bolivia), the coca market had an annual turnover of 1 million pesos, and food and clothing were sold for only 400 thousand pesos a year. But already in 1576, the Council in Rome banned the "devil's potion", and King Philip II of Spain issued harsh laws against the use of coca.

A few centuries later, coca re-emerged from oblivion in a rather unusual way: in 1863, M. Angelo Mariani created a new wine that became a resounding success. According to various accounts, this wine graced the tables of many famous and influential people of that era, including the Pope. Along with other ingredients, the wine contained coca leaf essence. It soon fell partly under prohibition, partly out of fashion and due to these circumstances was forgotten. But in 1885, a pharmacist from the United States, John Smith Pamberton, invented a headache remedy in which the main ingredients were alcohol and - again - coca leaves. Alcohol was quickly replaced by carbonated water, and soon the drink was being sold in many American pharmacies. Then a patent was bought for packing coca-cola into bottles and the drug began its march around the world as a soft drink. However, quite quickly the coca was removed from it, finally turning the drug into lemonade.

A year earlier than Pamberton's invention, in 1884, Sigmund Freud recommended that his patients take cocaine to relieve nervous tension and fatigue.

The "magic powder" quickly conquered the world market, and only a widespread discussion in the press led to its prohibition in 1912. In fact, cocaine can be recognized as the first chemical stimulant to appear on the market.

The history of drug stimulation

Man has far from always sought to give his thinking absolute clarity and precision. Very often, on the contrary, people have sought means to cloud their thinking, to disconnect from the surrounding stimuli and to reach some new level.

There is no point in talking about drugs, which, unfortunately, are becoming more and more popular nowadays; it is much more interesting to talk about where the tradition of drug stimulation and drug intoxication came from.

Opium poppy seeds are often found among food waste in Stone Age settlements in Europe. Remains of various narcotic plants are also found during excavations in other parts of the world. Many people immediately ask: if in ancient times there was no prohibition on drugs, why did not there was an outbreak of drug addiction and mankind did not die out before it was born?

Everything is elementary: life was not as simple as it is today, and a person who took a drug knew that under its influence he risked not seeing the abyss, getting lost in the forest, falling on the fang of a predatory animal or on the spear of an enemy. Natural selection was not dormant, and mass drug addiction was not destined to develop in the cruel primitive world. But already in the relatively prosperous times of antiquity Pliny the Elder (24-79 B.C.) never tired of warning his fellow citizens: "death awaits him who abuses opium".

Ordinary mortals had no access to drugs, and various herbs and seeds with special effects were used either by herbalists to cure the sick or by priests to communicate with the gods. It was observed that drugs not only gave a cheerful state of mind, but also had a stimulating effect. In ancient China, during the Yin period (XXI/XVI-II-XI centuries B.C.), Emperor Sheen Shung advised his subjects to take cannabis infusion as an invigorating tea and as a remedy for absent-mindedness.

In European history, the first crusades (X-XII centuries) and the voyage of the Venetian merchant Marco Polo (1271-1295) marked a turning point in the attitude towards drugs. It was then that hashish and opium came to Europe. The latter took a strong position among the widely used medicines.

Opium appeared in Russia in 1581, when the first royal pharmacy was opened in Moscow, run by Englishman James French. Opium was used mainly as an anesthetic or sleeping pill. But in the 19th century opium became a treatment for mental illness and alcoholism. Professor Spassky, for example, in 1883 recommended that "bitter drunks" should be used for white fever according to the method of Dr. Pauli, i.e. in the morning to give the patient a small shot of vodka, and at night one - two grams (one gram equals 64.8 mg) of opium in extract and rubbed with ointment, also infused with opium.

It is clear that very soon opiate addicts appeared and began to be treated with morphine extracted from opium, which, as it was believed, was not addictive. There were disputes about this throughout the XIX century, but in 1898, Charles Richet, a famous physician, wrote that "children do not form a habit to morphine, and small doses give a greater effect; in habitual users and colossal doses do not give a toxic effect.

However, cocaine soon became the cure for alcoholism, opiate addiction and morphinism. And then heroin, invented at the turn of the century, became the cure.

Hashish has not disappeared either. Many famous writers, poets, artists and journalists of the 19th century saw it as a powerful stimulant of creativity. Charles Baudelaire, Théophile Gautier, Alexandre Dumas-father, Gustave Flaubert were full members of the "Club of Hashishheads" that existed in Paris.

Already in the twentieth century, the drug researcher and famous writer Aldous Huxley wrote in Drugs That Shape People's Minds: "The problem of addiction to drugs and excessive drinking is not simply a matter of chemistry and psychopathology, of release from pain and conformity to a bad society. It is also a problem of metaphysics - a problem, almost, one might say, of theology." And he is right at least in that the effects of drugs on the brain were originally very closely linked, as we said above, to various religious cults.

By using different kinds of drugs, mostly psychedelics, the priests could talk to the gods. Or at least they thought they were doing so. The action of psychedelics is to turn off the filters that filter out details that the normal state of consciousness considers unnecessary (we have already talked about this in the chapter about geniuses and autistic people, when we discussed whether it is possible to find a button that will make a person a genius). The classic psychedelics are LSD and mescaline. The effects of mescaline, contained in the peyote cactus, were described in detail by Aldous Huxley, and the practice of its sacred use by Indian shamans was described in his books by Carlos Castaneda. Castaneda claimed that in 1960 he met the Indian sorcerer Don Juan and until the latter's death took lessons from him in magic based on the traditions of the Indians of ancient Mexico.

LSD was first produced in 1938 by the Swiss chemist Albert Hofmann, but its psychotropic properties were not discovered until 1943 by accident. Hofmann dreamed of treating schizophrenia with LSD, but many scientists did not believe that psychedelic and schizophrenic psychosis were identical. Hofmann's theory was disproved, but in the early 1950s, every significant psychiatric institution in the world was conducting experiments on humans and animals. However, many believe that the research was not followed through: LSD became too widespread among young people and was banned.

Now, scientific medical journals are increasingly raising the question of lifting the ban on LSD and resuming research. According to many researchers, among whom Stanislav Grof is best known, the use of LSD can lead to significant psychotherapeutic effects, relief of migraines, treatment of alcoholism and other positive therapeutic results.

Psychedelic substances can cause synesthesia (when stimulation of one sensory organ simultaneously produces sensations corresponding to another sensory organ, e.g.: colored hearing, colored sense of smell, rustling of odors), feelings of fear, depression and, almost simultaneously with it, euphoria, abrupt transitions from one state to another, auditory and visual hallucinations, vivid experiences of past events (especially from early childhood and infancy), temporal and spatial disturbance of world perception, illusory extraordinary abilities (e.g., the illusion of telepathy).

Although a number of Western universities and foundations continue to conduct research on psychedelics despite the ban, it is too early to draw any conclusions. Let's put it this way: if psychedelics are allowed, it will be our chance to discover another side of human consciousness.

William James, in The Varieties of Religious Experience, wrote: "The power of alcohol over mankind undoubtedly derives from its ability to stimulate the mystical properties of human nature, usually nailed to the ground by cold facts and dry criticism of the sober hour. Sobriety diminishes, disunites, and says "no." Intoxication enlarges, unites, and says yes... For a moment it makes one one with the truth. It is not out of mere perversity that men seek it...Intoxicated consciousness is only a piece of mystical consciousness, and our general opinion of it must find its place in our opinion of that larger whole."

And one more interesting quote: "Our normal or, as we call it, rational consciousness represents only one of the forms of consciousness, and other forms, completely different from it, exist next to it, separated from it only by a thin partition".

Whether science manages to get past that partition is up to the politicians.

CHAPTER IV. "MENTAL ILLNESSES" - OR HAS THE BRAIN SIMPLY TUNED IN "TO SOMEONE ELSE'S WAVE"?

One of the great hopes of the twentieth century, the cure of schizophrenia with psychedelics, has, alas, failed. The disease is still untreatable.

However, is schizophrenia a disease?

There is schizophrenic genius among the varieties of genius. In practice, it is regarded as a mental illness that can be transformed over time into a deep mental disorder. For example, many members of Charles Darwin's family had schizophrenia. And one of the greatest mathematicians of the twentieth century, winner of the 1994 Nobel Prize in Economics, John Forbes Nash, Jr. contracted schizophrenia in his thirties, was involuntarily committed to a private psychiatric hospital in suburban Boston, and subjected to chemical treatment. Released with the help of his lawyer, he traveled to Europe and attempted to seek political asylum there, but was extradited back to the United States. His condition worsened, and at times, one might say, Nash was not present at all in the world of reasonable people. But in the early 1980s, the mathematician's health began to improve, and today he considers himself almost completely healthy.

"A split mind."

Schizophrenia (from Dr. - Greek σχίζω - "splitting" + φρήν - "reason") is a mental disorder characterized by abnormalities in the perception of reality or its reflection. Symptoms are most often auditory hallucinations, paranoid or fantastic delusions, or disorganized speech and thinking against a background of significant social dysfunction or loss of performance. Among the pathogenic factors are genetic predisposition (Nash's son, by the way, is severely ill), early childhood living conditions, neurobiological disorders, difficulties in psychological or social interaction. But, in principle, the symptoms and causes of this disease are so different that some experts define schizophrenia as several different diseases, which we ignorantly reduce to one.

Diagnosis of schizophrenia is highly unreliable and inconsistent, and there is still no clear clinical test. David Rosenhan conducted a study in 1972 that resulted in an article in Science entitled "On Normal People in Abnormal Places," which proved that normal people in unusual circumstances behave almost identically to schizophrenics.

There has long been an "anti-psychiatry" movement, whose proponents peaked in the 1960s. Psychiatrist Thomas Sas believes that psychiatric patients are not sick, but rather are original individuals (with unconventional thoughts and behavior) who make society uncomfortable. In his opinion, there is no such thing as "schizophrenia" in principle, only a social construct based on society's ideas of normal or abnormal: "Schizophrenia is so vaguely defined that in reality the term is often applied to almost any kind of behavior that is not to the **liking of** others".

Sas, unlike the "anti-psychiatrists," is not opposed to treatment for schizophrenia, but emphasizes that it should only be given with the consent of the patient. This view is shared by a number of prominent Western psychiatrists who believe that the symptoms of what is commonly referred to as mental illness are often *reactions to the impossible demands imposed by social or family life on some hypersensitive individuals*.

In 1956, a theory of schizophrenia emerged in which the onset of this disorder was attributed to the consequences of a person's exposure to different or contradictory information. Julian Jaynes, whom we have already mentioned at the beginning of our book, in his work "The Origins of Consciousness in the Destruction of the Two-Chambered Psyche" suggested that before the beginning of historical time schizophrenia or some similar condition was the norm of human consciousness. According to his theory of the "two-chambered mind," a state of low affect suitable for routine actions may be interrupted in moments of crisis by the appearance of "mysterious voices" giving instructions, which in ancient times people considered to be the advice of the gods.

To this day, many tribes choose as shamans people who, from the medical point of view, are sick with schizophrenia: they hear voices, travel to parallel worlds, and so on. However, according to some researchers, these people are not sick, but can simply induce states that give a full clinical picture of schizophrenia.

If this is true, we can assume that it is possible for a healthy person to "switch" their brain to a schizophrenic perception of reality and then return to normal. So, we can assume that it is possible to "switch" the brain of a schizophrenic sufferer in a similar way. Consequently, it is also possible to switch from normal functioning of consciousness to "genius" consciousness (it is possible that the abilities of shamans may become the key to the treatment of schizophrenia!).

The term "schizophrenia" is very often misinterpreted as "split personality".

Although some patients hear voices and even perceive them as separate personalities, **there is no change of personality in** schizophrenia. This is quite rare and scientifically curious.

The five "I's" in Ms. Fisher's body

Many psychologists believe that there are at least two personalities in the mind of each of us. Anyone has been told, "it's like you've been switched" or "it's like you have two different people living in you". Although such phrases are usually uttered without any backward thought, they are much closer to reality than one might assume.

We remember that the cerebral cortex is divided into two hemispheres, the functions of which are very different. The right hemisphere is responsible for figurative thinking, while the left hemisphere is responsible for logical thinking. In fact, these are two personalities - the cynic and the romantic, the dreamer and the pragmatist.

Few people have both of these personalities in perfect balance, usually one of them is repressed and is only released when the other has weakened its control for some reason. For example, a successful and punchy person, when his main personality sleeps, in dreams can suffer from failures and failures, fears and complexes. Or the kindest and most intelligent person, it is worth it thoroughly "too much", turns into an insolent malicious boor. Each of us has probably encountered something similar or even observed such transformations.

But it happens that no reasons are necessary for the manifestation of inner personalities. For example, Helena Petrovna Blavatsky (1831-1891), who laid the foundations of Theosophy and wrote many books, believed that some higher beings, mahatmas, dictated these books to her consciousness. In a letter to her friend N. A. Fadeeva Blavatsky said: "Think only that I, who studied exactly nothing in life, I, who neither about chemistry, nor about physics, nor about zoology - as it is, had no idea - now write dissertations about all this. I enter into disputations with scientists and come out victorious..." In Blavatsky, according to her, the most different personalities "entered". One of these personalities was a brilliant stylist, an absolutely literate mahatma with calligraphic handwriting. Another personality wrote untidily, with many mistakes, forcing Blavatsky to behave in a bad way - to swear foully, to show intolerance and so on. Naturally, such things have a heavy impact on mental health, and Blavatsky suffered from mental illness, which, however, did not prevent her from working fruitfully for the rest of her life.

And the most famous classic case of split personality is the story of Miss Christine Besham. In the body, or rather, in the brain of this girl lived four separate personalities, which differed from each other not only in character, but even in health, level of knowledge and memories. They were "nasty Sally," "Besham the fearful," "Besham the sincere," and "Margaret." Sally had a knack for hypnotizing other selves, controlling them and often tormenting them. Putting toads and spiders in various places, for example, so that the other Ms. Beshams, suddenly finding them, experienced a few unpleasant minutes. Or, for example, she once took the last bus out of town, got off at the last stop, and then "went away," leaving in her place a "fearful Besham," who, trembling, had to walk back to town.

Dr. Morton Prince, the psychiatrist who observed Miss Besham and described her in the medical literature, had long tried to merge the four personalities into one, but this was stubbornly opposed by Sally, saying that she was a spirit and could not merge with anyone. The doctor had to "banish" Sally, and then the three remaining personalities merged into one Miss Besham.

Another such case is that of Doris Fisher, who had five separate selves, known as "True Doris," "Sick Doris," "Sleepy Doris," "Margaret," and "Sleepy Margaret. It was "Margaret" who was the prankster - she liked to steal something, setting it up so that "True Doris" would be suspected of stealing it. She would wear her best dress and jump into the river, or she would brush her body with blood so that only "True Doris" would feel the pain.

The doctor - Walter Franklin Prince, the namesake of his colleague, trying to understand this story, came to the conclusion that this case did not resemble "projections from the subconscious" at all. He decided that there was a "non-physical" entity from the outside that had subjugated Doris Fisher's own self.

Prince turned to James Hislop, a professor of logic and ethics at Columbia University who had devoted much time to the study of hysteria, split personalities and the like. He too could do nothing, and then a medium was invited. Strangely enough, with his help, the girl was cured. The "foreign entity" was banished.

"Doppelgangers" can live in the same body for years and decades, and in many cases for the entire life of the patient. Interestingly, there is always a main "doppelganger", but this role passes from one personality to another, often through intrigue and other "warfare".

Role reversal usually occurs when a person faints, and upon waking up, he or she becomes a different person, usually without remembering his or her previous "incarnation" and often without even realizing that several personalities coexist in him or her.

How not to remember computer terminology again - it was just a "reboot of the system under another user". And the new user cannot find out what was going on with the same computer five minutes ago, what programs were launched, what was changed... Imagine how the brain works in such cases: a person is, if in such a case one can say so, sane and sober, can hold a conversation, navigate in time and space, do some work... But only that work, which is trained by the personality, which at the moment controls the body. All his memories, habits, skills belong to the "awake" personality, and he may not have the slightest idea about the rest.

The device of a real computer is well known, but how the computer-brain works, alas, it is not clear, and the essence of the phenomenon of "doppelgangers" has not yet been solved.

Personality variations

Medical doctors believe that in split personality the integrity of the human "I" under the influence of emotional stress or other reasons is broken into pieces. The person as if says: "I did not do it", "it did not happen to me". And the fault line runs along the most vulnerable places, which can be, for example, unrealized ambitions, long suppressed desires, childhood traumas, and so on. And then a person in whom each trait was balanced by the opposite one, for example, generosity - greed and so on, turns into several personalities in which these character traits are deprived of counterbalances.

Bifurcated personality disorder is known scientifically as dissociative identity disorder. Its diagnosis requires the "presence" in the patient's psyche of at least two personalities, each of which regularly takes turns controlling the individual's behavior, and memory loss beyond normal forgetfulness, referred to by the term "switching." These symptoms are usually completely independent of any substance abuse (alcohol or drugs) or general medical condition.

Many patients with dissociative disorder try to hide their symptoms (and the illness usually begins in childhood), and so some individuals remain children. The first medical observation of such a disorder is Paracelsus, who in 1646 described the case of an anonymous woman who claimed that someone was stealing money from her. The thief, as it turned out, was her second personality, whose actions the first personality did not remember.

In 1784, the Marquis de Puissegur, a student of Franz Anton Mesmer, one of the founders of hypnosis, hypnotized his employee Victor Ras into a somnambulic state, but Victor remained awake during sleep. Upon awakening, he did not remember what he had done in the altered state of consciousness, whereas in the latter state he retained full memory of the events that had happened to him in both of his hypostases.

In 1791, Eberhard Gmelin described the case of a 21-year-old German girl who manifested a second personality who spoke French and claimed to be a hereditary French aristocrat. The phenomenon of this case was that the German herself did not know French.

Such cases have been described by physicians on numerous occasions. In 1888, doctors Burroux and Burro published a book entitled "Personality Variations", where they told about a certain Louis Vivet, in whom six different personalities coexisted. Each of them had their own individual memories, hard-wired to a specific period of Louis' life. Based on these observations, researcher Pierre Janet introduced the concept of "dissociation" and suggested that these personalities were coexisting psychic centers in the body of one individual.

In 1899, an even more interesting case was described. Theodore Flournoy in his book "From India to Planet Mars: A Case of Somnambulism with Fictitious Languages" described individuals who spoke unknown languages (remember the genius autist who created full artificial languages).

In 1954, Thigpen and Cleckley's book The Three Faces of Eve was published, which artistically conceptualized the story of psychotherapy involving Kris Costner-Sizemore, a patient diagnosed with multiple personality. In 1957, a movie starring Joanne Woodward was made, and in 1977, Kris Costner-Sizemore herself released an autobiography, I Am Eve, in which she claimed that Thigpen and Cleckley's book took great liberties with her life story.

Subsequently, other works on multiple personalities have been published. In 1980, the book "Michelle Remembers" was published, authored by psychiatrist Lawrence Puzder and Michelle Smith - his patient. A year later, those interested could read "The Multiple Minds of Billy Milligan", compiled by Daniel Keyes from interviews with Billy Milligan and his psychotherapist.

The last decade of the century was revolutionary: in 1995, Astraea\'s Web site appeared, dedicated to the idea of recognizing multiple personality as a healthy condition. Three years later, the famous magazine "New Yorker" published an article by Joan Akokella "Creating Hysteria", which talked about medical "excesses" in the treatment of multiple personality. In 1999, Cameron West's book First Person Plural: My Life in All Its Diversity was published, and in 2005,

Robert Oxnam's The Split Mind was published. Both books were authored by people with multiple personalities.

There are more and more articles in medical publications that multiple personalities do not necessarily need to be treated.

Interestingly, a count conducted in 1944 found only 76 described cases of multiple personality in the nineteenth and twentieth century medical literature. Many physicians in the thirties and forties were even of the opinion that the disease was a thing of the past, as no patients had been recorded for several decades. Why this happened - no one could explain, although various theories were put forward. However, the disease has returned again on a much larger scale, with around 40,000 cases of multiple personality being reported between 1985 and 1995. To explain this sudden outbreak, again, no one has yet been able to explain it.

Lost souls

If scientists are lost in speculation, mystics are firmly convinced that the phenomenon of doppelgangers is the fault of the souls of the dead, who entered a random body, whose defense was weakened. In Christianity, however, it is believed that not souls but demons, and there are even special rituals of scrubbing and exorcism to expel uninvited guests.

The case of a five-year-old Indian boy, Titu Prasan, who told his parents that he was one Suresh Varma, a shopkeeper in Agra, had a wife, Uma, and two children, and that he had been shot in the head. At first the parents took it as a childish fantasy, but the boy told more and more details: he allegedly owned a radio store, and at the age of thirty he was killed by two shots while driving a car to his house....

Often after such stories, the boy would become hysterical, throwing objects at his family and shouting that they were not his parents. The joke, if I may say so, was getting long and Titu's parents, Shakti and Mahavar Prasan, decided to check what he was saying. To their utter amazement, it turned out that a radio store owner had indeed been murdered in Agra five years ago. The boy's parents called the widow and told her about their child's surprising behavior. Uma along with the parents of the late Suresh decided to visit the strange boy and as soon as their car reached the Prasan house, Titu immediately rushed to hug his 'wife' and his 'parents'. He not only recognized them and called them by their names, but was also upset that Uma had come in a Maruti car and not his old Fiat, which, as it turned out, had been sold.

The boy managed to answer all the tricky questions that the "first" family asked him. This case received a lot of publicity, hit all the media of the world, and even in the USSR the newspaper "Trud" wrote about Tita. Naturally, scientists became interested in such transmigration of souls. N.K. Chanda of Delhi University and E. Mills from the University of Virginia (USA) once again carefully questioned the boy and his relatives, asking them in parallel questions about various unimportant events. But the boy passed this exam with honor. During the examination of Titou scientists found a strange skin scar on his right temple and a birthmark over his right ear. The autopsy report on Suresh Varma's body stated that the bullet had hit him in the right temple and ricocheted off the skull and exited above his right ear.

This case of the transmigration of souls, which could be considered virtually proven, aroused great interest in India, a country where reincarnation is traditionally believed in.

Experts from the Bangalore Institute of Mental Health and Neuropathology conducted a study of two hundred and fifty cases like Titu's that have been reported since 1975. During the ten years that the study lasted, its initiator, Dr. Satwash Pasricha, personally traveled to many villages and interviewed at least twenty eyewitnesses for each such case. Many cases were not confirmed, but those that the scientists found to be quite credible were enough for the following scientific conclusions.

"Reincarnation effect" is observed mainly in children between the ages of three and seven. Over the years, they completely forget about their "previous incarnations". "Reincarnation" in a child is usually caused by a random remark of someone from the family, which reminds him of the events of his past life. In 82 percent of cases, the children clearly recalled their name in the "past birth" and details of that life, which were enough to make an identification with the deceased person. Interestingly, more than 50 percent of the reincarnated souls had died violent deaths in their past lives, and their average age was about thirty-four years. In 80 percent of cases, the children were not related to the deceased by even distant ties of kinship, which rules out the phenomenon of "genetic memory."

Interestingly, the belief in reincarnation existed in early Christianity, but was abolished in 553 by the Second Council of Constantinople.

Stranger wave

However, many researchers believe that the transmigration of souls has nothing to do with it, and the so-called brain waves are to blame. Talks about this phenomenon began in the XVIII century, but until now it was not recognized by science, and the propagandists of this idea was considered a fraud and pseudo-scientists. However, in 1849, Du Reymond proved that the brain, as well as the nerve and muscle, has electrogenic properties.

On August 24, 1875, English physician Richard Caton presented data in a report to the British Medical Association that the brains of rabbits and monkeys emit weak electrical impulses. And in the same year Russian physiologist V. Y. Danilevsky in his doctoral dissertation outlined the data obtained by studying the electrical activity of the brain of dogs. And in 1882, the famous I. M. Sechenov published the work "Galvanic phenomena on the medulla oblongata of the frog", in which he proved the fact of rhythmic electrical activity of the brain. In 1884, N. E. Vvedensky listened to the activity of the frog's medulla oblongata and rabbit's cerebral cortex with the help of a telephone. He confirmed Sechenov's basic observations and proved that spontaneous rhythmic activity can be found in the cortex of the large hemispheres of mammals.

And in 1913, V. V. Pravdich-Neminsky published the first electroencephalogram recorded from a dog's brain and introduced the term "electrocerebrogram", which is unfamiliar to us. The first human EEG recording was obtained by Austrian psychiatrist Hans Berger in 1928. He proposed to call the recording of brain biocurrents "electroencephalogram".

Today, EEG registration has become a routine procedure, which is performed with special electrodes, each of which is connected to an amplifier, and they are placed in a certain order on the head of the person under study.

"Brain waves" are low-intensity electromagnetic waves emitted by the brain, with frequencies ranging from 1 to 40 hertz. Five major groups of brain waves are distinguished:

- delta waves (0.5-3 Hz) - they appear during deep sleep, trance or hypnosis;

- theta waves (4-7 Hz) - emitted during sleep, deep relaxation and meditation;

- alpha waves (8 - 12 Hz) - indicate that the person is in a borderline state between sleep and wakefulness or in meditation;

- beta waves (13-30 Hz) manifest themselves in an active, alert state. Their high activity corresponds to a high release of stress hormones;

- gamma waves (30 Hz and above) - according to many reputable scientists, are a sign of "hyperconsciousness".

Today, it is practiced to externally stimulate the brain with waves of the same frequencies that it emits on its own. This puts our brain matter into a state in which waves of a given frequency would occur naturally. For example, an extremely agitated person can be relaxed by stimulating their brain with alpha waves for five minutes. And theta waves cause positive emotions, a feeling of comfort and harmony. They are used in various "high-speed" methods of audio or video training. It has been proven that simply reading a textbook under theta waves helps to better assimilate the material.

Brainwave synchronization is now widely used in the West to treat depression, migraines and headaches, autism, absent-mindedness, alcohol and drug addiction. Unfortunately, in our country this branch of medicine is considered almost quackery, although in other countries the works in this field have taken a place in solid scientific publications.

In general, interest in the topic of brainwaves is quite high in the West, and recently there has even been a computer game that controls the brain without a mouse and keyboard. The Mind Balance game presented a few years ago at the MIT Media Lab Europe in Dublin is rather primitive, but the first step in this field has already been made, and perhaps in a few years we will be able to control thought with the help of computer programs.

Before starting the game, a person wears headphones with six attached sensors that are fixed at specific points on the back of the head. These points correspond to areas of the brain responsible for processing light signals, vision and even hallucinations. The captured electroencephalogram is transmitted to a computer using Bluetooth wireless technology.

The essence of the game is very simple: a character walks along a rope, and the player has to help him keep his balance: if he leans to the left, look at the right side of the screen, and vice versa. It is clear that so far it is more like entertainment, but perhaps in time it will be invaluable help in the development of more perfect interfaces for communication of completely paralyzed people whose brains continue to function perfectly. Pilots, astronauts and many other people in active professions will also benefit from this invention.

And in the Netherlands, a device has appeared that is able to distinguish between sounds heard by a person by scanning the brain. Each sound leaves a characteristic "neural imprint" on the auditory zone of the listener's cerebral cortex. It was recorded in the following way: the subjects were pronounced vowel sounds, and the computer, taking their EEG, decoded these signals. So far, scientists can recognize only individual sounds, but soon, as the head of the research group Elaya

Formisano from the University of Masstricht, the machine will be able to fully understand what a person hears, just scanning his brain. And then, we would add, it won't be long before it can read minds.

And one more fact, already rather from the field of curiosities (although, of course, what the hell is not joking). "Journal of Applied Psychology" published a comparative analysis of encephalograms of music students and non-musicians. In creative natures was found a good synchronization of biopotentials in different parts of the brain, the predominance of right hemisphere activity and high frequency in the delta rhythm range. Moreover, medics have found that the encephalogram can catch differences in the degree of giftedness, and now they propose to make EEG an additional tool in the selection of applicants for the exam.

We wrote above about chemical brain stimulants and the fact that employers may be obliged to take them, but now it turns out that along with lie detectors, which are used for many positions, employers may soon use EEG to select the most talented applicants. In fact, it will turn out that the skills and abilities acquired through years of study and labor may not be used, if God or nature did not initially give a person talent. Fortunately, such technologies do not exist yet, but it seems that we are on the threshold of the dark times that Hollywood sci-fi writers are so fond of telling us about. And there is no doubt that, putting aside the imperfect lie detector, the employer will be a direct mind reader....

But back to brain waves. If they can be captured and stimulated with the help of a mechanical device, then it is logical to assume that the brain of each of us easily receives the waves of our loved ones or even those with whom we sit next to in the same office or just shaking in public transportation. It is not yet known whether it decodes them or puts some kind of block in their path. But some experts believe that many so-called mental illnesses come from the fact that something breaks in the brain and it begins to perceive other people's brainwaves as its own.

Is it possible that the delusions, fears and visions of schizophrenics are just someone else's brainwaves caught in space by the patient?

It is known that our planet emits waves at approximately the same frequency as the human brain. We do not know what information these waves carry, what they can tell the brain "off course", we do not know. Therefore, it is quite possible that the case of the Indian boy is not a story about the transmigration of souls at all, but simply a story about the fact that the child's brain, which has not yet fully mastered its functions, accepted the wave of horror released by a man at the moment when the killers shot him with a gun. And the child's memories were not formed independently, but were entirely substituted by the experience of the owner of the radio store.

There are documented stories that can only be explained by a similar theory of the brain's reception of other people's waves.

In 1978, Nikolai Alexandrovich L., a resident of the Vologda region, was struck by lightning. He not only survived, which, in principle, sometimes happens, but also spoke three European languages, which he had never studied, quite fluently.

In 1987, Gennady Sergeevich S., a resident of the Tula region, was pinned against a fence by a truck trailer. Pensioner S. received, among other injuries, a severe head contusion, and waking up the next day, was surprised to realize that although he had not forgotten his native Russian, he found it much easier and more comfortable to express his thoughts in German. It was a language he had never known before.

Extraordinary abilities caused by a sudden effect on brain waves are not only about languages. Many suspect that the genius of little Mozart arose for some similar reason: it is known that Amadeus, for example, suddenly began to play the violin, although no one taught him to do so, and the violin is one of the most difficult instruments. And in general, how can you imagine a four-year-old child writing down the notes of the genius music he composed? The clergy, by the way, saw in the talents of little Amadeus the finger of God. I wonder what scientists of the XXII century, for example, would say about this?

Let's hope that science will someday still be able to solve the mystery of brainwaves.

CHAPTER V. THE SLEEP OF THE MIND. WHAT THE BRAIN DOES WHILE WE SLEEP. ARE PROPHETIC DREAMS A REALITY?

For as long as mankind has existed, it has wondered what sleep is. Aristotle believed that it was a state on the verge of life and death; it was not without reason that the ancient Greek god of sleep, Morpheus, was considered the brother of Death.

Many mystics were convinced that during sleep the soul travels through the spaces of the worlds, and believed that if a person is suddenly awakened, the soul will get lost before returning. The ancient Indians had a belief that it was impossible to paint a person's face during sleep, otherwise the returning soul might not recognize its body.

It was not until the discoveries of Darwin and Freud that sleep began to be understood as one of the mechanisms of brain function. But no matter how much scientists tried to reduce sleep to only physiological functions, they failed. People continue to believe in the mystical power of dreams and sometimes discover things in their dreams that modern science cannot explain.

In the realm of Morpheus

Arm yourself with patience - we will have to start with the basics of dream science. A person's night's sleep is usually divided into four or five cycles of 90 to 120 minutes (the duration of each subsequent cycle increases).

All cycles begin with the so-called slow or deep sleep (it is in this phase that manifestations of sleepwalking are observed), and end with a 10-15 minute phase of sleep with dreams - the so-called rapid or paradoxical sleep (by the way, the best remembered dreams that you see during the last cycles, from the sixth to the eighth hour of sleep).

During REM sleep, all signals from the outside world are no longer received and analyzed. It is difficult to wake up someone who is at the peak of the REM phase. At this time, the person's muscles are completely relaxed and the brain is working intensively. It is in this phase that the eyeballs move and the person can talk in sleep.

Incomprehensibly, a person starts dreaming while still in the womb. The embryo is in a state of so-called activated sleep. While in the mother's belly, the fetus moves, pushes, and all these are the consequences of the dreams it has seen. What can someone who has never seen the world dream about? It is not known. Some scientists believe (and there is some scientific evidence for this) that the dreams of the mother and the future child are synchronized, and he, seeing the dreams of his mother, in such an unusual way familiarizes himself with the world. How "pictures" are transmitted from one consciousness to another is a mystery and undeniable evidence that we know too little about the principles of the brain. Are shared dreams transmitted through blood or through bio-waves unknown to us yet? The answer to this question has not yet been found.

We spend a third of our lives, if not more, in sleep. Of the sixty years we have lived, for example, twenty are spent sleeping. Sleep is rest, as we were taught in childhood. And indeed, the vital processes of the whole organism slow down, but the brain continues to work intensively, distributing the day's sensations and the information received to its "storehouses".

In the first 24 hours of a child's life, he sleeps for 22 hours and 80% of his sleep is "REM sleep", the phase during which we dream. In fact, the child has a lot of new sensations that need to be sorted.

With age, unfortunately, this phase of sleep becomes shorter and shorter. In a five-year-old child, REM sleep already takes about five hours, and in a thirty-year-old person who sleeps about eight hours a night, the REM phase, as we have already mentioned, is about two hours. By the age of sixty-five it decreases to one and a half hours.

Comparing the work of the brain to the work of a computer, we can say that while we sleep, the brain "rewrites" the information we have received from "RAM" to "hard disk", i.e. from short-term memory to long-term memory. All this happens during so-called REM sleep. Experiments on rats have shown that the sleeping animal has seven times stronger nerve impulses in the brain during this period than in the awake. Man, in this sense, has not gone far from the rats. It is during sleep there are processes associated with the creation of memories, or, scientifically speaking, "memory consolidation". A small part of the brain - the hippocampus - is responsible for this, and if it "breaks down", a person loses the ability to memorize anything: only old memories remain, which have already been "recorded" in the cerebral cortex.

It has already been proven that a **person who falls asleep after learning new material wakes up and understands it much better**, because during sleep the brain has systematized the information received. Scientists from the University of Chicago conducted an experiment: a group of people were explained the method of solving logical problems. After sleep the speed of solving these problems increased in the subjects by twenty percent. That is, during sleep the brain not only puts memory in order, but also develops, based on the experience gained during waking hours, the programs of our behavior.

The brain in general knows much more than we think. Dream specialists from University College London decided to figure out whether the brain can distinguish between bad and good omens. Volunteers were shown random pictures, certain combinations of which were accompanied by a strong electrical discharge in the brush. After a while, scientists noticed that the brain begins to react to these combinations before the electric discharge, and in fact the volunteers themselves had difficulty explaining the logical principle by which the pictures were combined with the discharge. Conscious recognition of such "bad" combinations began to occur only after the volunteers spent several hours in sleep.

Our dreams are just a shadow of the intensive work our brains are doing. Each of us, having fallen asleep for just a minute or even a few seconds, has noticed that during this time he or she has managed to see a rather large "slice of life" in a dream. At this time, the brain is not constrained by "behavioral limitations" and, no longer dependent on the body, it tries to reach its optimal speed of work 5 to 20 times faster than what we are used to when we are awake.

Scientific discoveries in a dream

Very often the solution to problems that we have struggled over in vain during waking hours comes to us in dreams. Perhaps the most famous examples are Mendeleev with his famous table and Friedrich August Kekule, who saw the structure of the benzene molecule in a dream in the form of a snake coiled in a ring and beating its tail. The brain, having accelerated, solved in a dream the question that the scientist could not answer for months. And "showed" this solution, though in a symbolic form, to its "owner".

But perhaps the most interesting story of "sleep discovery" happened to Dr. Otto Levy, a pharmacologist and Nobel Prize winner. He spent years studying the chemical transmission of nerve impulses. Levy couldn't solve one problem for a long time, and his research came to a halt. Suddenly he began to dream the necessary experiment. The first two nights Levy woke up and wrote down the necessary data in a notebook, but in the morning he could not make sense of his notes. The third time he did not write anything down, but went to the laboratory and simply put the dream experiment (which turned out to be successful), as he saw in the dream. Levy later admitted that if the idea had occurred to him while he was awake, he would have dismissed it as utter nonsense.

However, even in science sometimes mysticism interferes. The great Russian poet Mikhail Lermontov, as you know, liked to do math at leisure. Once, having failed to cope with a difficult task, he saw in a dream a stranger who dictated the solution to him. Lermontov not only wrote down this astounding dream, perceived by him as absolutely real, but also drew a portrait of a stranger. Many years later, researchers discovered that the portrait depicted a man very similar to John Neper - the creator of logarithms, who died two hundred years before this dream. Interestingly, for the mystic and interpreter of prophecies, Neper, math was also just a hobby. Incidentally, he was Scottish (Lermontov also had Scottish roots). Perhaps in some knee their genealogical trees crossed. It is quite reasonable to interpret the solution of the problem that came to Lermontov as the work of his brain, but how to explain the appearance of a mathematician unknown to

Lermontov in this dream? Today, researchers are quite certain that Lermontov could not have known who Neper was, much less seen his portrait.

Mikhail Yurievich, by the way, paid much attention to dreams. His poem "Dream", in which he predicted his death, and the vision of this tragic event to his beloved in St. Petersburg, later found confirmation in the memoirs of Evdokia Rostopchina....

We'll talk more about prophetic dreams below, but for now let's get back to the "science of dreams".

It has already been mentioned that the rapid phase of daily sleep takes a total of about two hours, and the rest of the time we sleep without dreams. But nevertheless, over the course of a lifetime, these short periods add up to significant numbers. The famous philosopher Arthur Schopenhauer wrote: "<...> the world <...> must be recognized as related to the dream or even belonging to the same class of things as it. For that function of the brain, which during sleep by some charms generates a completely objective, visual, even tangible world, must take the same part in the creation of the objective world of wakefulness".

Scientists have never fully understood how our memory works and why we dream. Specialists also find it difficult to say why the brain does not process information in parallel with its reception. Here, perhaps, is to blame the diurnal cycle of human life, formed by the alternation of light and dark periods on our planet. For now, it is worth simply recognizing that the work of the brain in sleep is the most optimal solution for our lives. When processing information, the brain turns off those sections from which it currently draws knowledge received in the waking state. It is clear that this is most conveniently done in sleep.

Academician Bekhterev once said that a dream is an unprecedented combination of former events. But how do dreams show us the future?

Prophetic dreams will warn you of illness

Prophetic dreams, foreshadowing various diseases, scientists have long since not considered something unusual or fantastic. Disease, as we know, does not begin suddenly, and pathological changes that you have not yet noticed, your body feels immediately: the sick organ begins to slowly malfunction and the subcortex, disinhibited during sleep, reacts to the signals given to it by the body. So, both the general physical condition of the body and the composition of the blood can largely dictate the content of our dreams.

Today, for example, it is proved that our dreams begin to warn us about Botkin's disease (jaundice) a week in advance, about neurosis - several months in advance. Those who suffer from chronic gastritis get the "news" a month before the onset of symptoms, about hypertension - 2-3 months, and the possibility of a brain tumor dreams begin to "sound the alarm" a year before the clinical manifestations of the disease! Even a sudden, seemingly sudden flu can be predicted in two days!

Doctors advise: if you persistently dream that someone chokes you or you get into a confined space - check your **throat**; such dreams can also be a sign of the beginning of bronchial asthma.

- If someone in a dream traumatizes your stomach (beats on it, sticks a knife in it) - it is necessary to conduct a study of the digestive organs. It is also believed that the brain signals an impending **stomach** disease with dreams of spoiled, rotten food, visions of your own or other people's vomit.

- In **neurosis,** people dream of aggression, their dreams do not have a logical conclusion or end badly. The action of such dreams often takes place in an unfamiliar environment, and the people the neurotic sees in the dream are unknown to him.

- **Heart** problems are indicated by experiencing intense fear (including fear of death) in a dream, as well as being wounded or stabbed in the chest.

- Dreams of choking, tight clothing that makes it difficult to breathe, or squeezing through narrow gaps can signal **lung** disease.

- A dream with insects crawling all over your body hints at **skin** diseases.

- The impending **cold** or **flu** portends a dream with bathing in icy water or snow, the feeling that you are freezing.

- Dreams with uncomfortable hairstyles or tight headgear speak of head diseases.

But the main condition that these dreams, these images indicate this or that disease is their repeated repetition in your dreams for some time or several days in a row (as in the case of the flu). In addition, if in a dream a person's heart rate increases, he wakes up in a cold sweat, feels numbness and coldness of the extremities or "goosebumps on the skin" and it is repeated repeated repeatedly - it is necessary to undergo a general medical examination. Most likely, this is the body's way of warning about the development of a serious disease.

The famous American paleontologist Sternberg believed in prophetic dreams. He first encountered mysticism in this way. One of the museums ordered him leaves of ancient plants, and he spent the whole day thinking about where to get them, and then he saw in a dream that the plant he needed was at the foot of a mountain. The name of the mountain was known to him, but he had never been there. Intrigued by the "all too real dream", Sternberg went to the place he had seen in his dream in the morning and found what he was looking for.

Since then, the study of dreams became his hobby. He believed that if the body reports an impending illness in a dream, it is possible to program the body to fight it in the same dream.

Biologist Berg, who has done considerable research in this area, reported that one of his colleague's patients, a stomach cancer patient, tried to use the technology of conscious dreaming to find a healer in the dream world. Two weeks later, she dreamed of a man who seemed quite powerful. She approached him and told him about her illness, asking for help. He took her to a brightly lit mansion where there were many people and some kind of celebration was going on. She was placed on a couch and some liquid was poured on her stomach, after which she was told that the disease had been stopped. After a while, the doctors did indeed find that the cancer had stopped developing, and a couple years later the woman recovered without any medical intervention. The powerful man came to her several more times in her dreams, giving her advice and carrying out some manipulations. It would be possible to attribute this story to the biological tuning of the body to heal, but Berg, interested in the unusual dreams of the patient, conducted additional research and found that the powerful character of her dreams came in dreams and several other people who were cured miraculously.

"As a scientist," Berg wrote, "I am too far removed from mysticism to attribute these dream adventures to the visitations of some angel or saint. Perhaps it can be said that the subconscious mind of all people works in approximately the same way and we see in our dreams in one way or another similar things. But if we take this theory as a basis then there will be even more contradictions. And so I'll just say that modern science has no idea how to explain these documented cases".

Trust in prophetic dreams.....

However, this is where science's recognition of prophetic dreams ends. Amazing dreams, in which people very clearly saw the coming events, science either denies at all, or refers to the category of accidents. But many people will not agree with this position. Prophetic dreams have been known to man since antiquity. Calpurnia, the wife of Julius Caesar, in a dream saw that her husband will be killed, and persuaded him not to go to the Senate, in detail telling him about the dream. Brutus, a friend of Julius, who was present, said that Caesar should not believe in "woman's dreams" ....

Emperor Augustus, unlike Julius, listened to his friend's dream and did not sleep in his tent. On the same night, the enemies who sneaked into it in the darkness chopped up the Emperor's empty bed with swords.

There is a well-known story that took place in the nineteenth century with Captain Jonathan of California. He dreamed of a group of exhausted people dying of hunger and cold in the mountains. In the morning he told his dream in detail to a friend, a hunter and expert in those places. He recognized Carson Valley from the description. The captain, who had never been to this gorge, which was 150 miles from his home, risked organizing an expedition with his own money and managed to rescue several people lost in the mountains.

It is known that eighteen people gave up their tickets for the legendary Titanic flight. As it turned out later, the reason for the refusal of the flight of five of them were dreams, seen either by themselves or their loved ones, in which, one way or another, figured the disaster of the super-ship. Coincidence?

French dream researcher Camille Flammarion wrote: "I strongly assert that the phenomenon of dreams accurately predicting future events must be recognized as an undoubted fact... We can see without the aid of our eyes and hear without the aid of our ears by means of some inner mental and intellectual sense. Through inner sight the soul is able to see not only what happens at a distance, but can also know in advance what is to happen in the future. The future exists potentially, conditioned by the causes causing subsequent events."

You can certainly argue with Flammarion, but haven't you noticed that if, for example, you stare intently at a person who is, say, on the other side of the street, he will soon turn around and search with his eyes for the person who is looking at him? The phrase "felt someone's gaze" does not seem unusual or invented to us. But in the meantime, this feeling can not be supported by any scientific justification. Even the term "coincidence", so beloved by scientists as an explanation of incomprehensible phenomena, is not suitable. We know from ourselves that we are able to feel someone else's gaze and can feel that our children are in trouble. And many other things that modern science does not find an explanation for. But still do not fall into mysticism: our brain has much more capabilities than we think. And these abilities are still very little researched. However, some scientific institutions are studying such phenomena, and it is possible that soon our ideas about our own abilities will be greatly expanded.

It is known, for example, the record of the mother of the Decembrist Kondraty Ryleyev, where she tells about her prophetic dream, which came true only after many years. When little Kondraty, or Konichka, as he was then called, fell ill and the doctors agreed that there was no hope, his mother, Anastasia Matveyevna, praying at her son's crib, fell asleep, and saw in her dream an angel who warned her that it was better for Kondraty to die now. She began to beg for life for her son, and then the angel led her through an enfilade of rooms, asking each time before opening another door: "Have you changed your mind yet? When the door is opened - the decision will not be changed." Consistently Anastasia Matveyevna saw her son's entire future life. In one of the last rooms it was very smoky. Some young people were arguing there, and the woman realized that she saw a conspiracy being prepared, but she continued to insist. Then the angel opened the last door and the mother saw her son on the gallows. And this future could no longer be prevented.

Lincoln's dream, about which he told his relatives, is also known. A few days before his death, the American president dreamed of a completely empty White House, and he wandered around in amazement until he realized that all the people were crowded in one of the corner offices. He went in there, and when he saw the coffin, he asked the soldier standing guard, "What's going on?" The man replied, "The president has been killed in the theater." A few days later the dream came true.

Lydia Vertinskaya, wife of the famous chansonnier, recalled that immediately after the death of her husband, she dreamed that she saw her husband off at the station and he said to her: "Look, do not forget - I have the fifth carriage, the sixth seat." "The next day I went to get a pass to the Novodevichy Cemetery, where Alexander Nikolayevich is buried. In the pass it was marked: 5th section, 6th place."

These instances can no longer be explained by coincidence or chance, and we can only recognize that there is a certain mystical level that we can only reach so far in dreams.

By the way, Academician Bekhterev, a prominent Russian medical psychiatrist, was very attentive to prophetic dreams and in no case considered them an accident. In 1910 he instructed the practicing physician P. I. Vinogradov to find out from his patients whether they had ever seen any prophetic dreams. Vinogradov from 1910 to 1914 collected more than one and a half thousand testimonies, from which it followed that at least once in his life, a prophetic dream has seen almost every second interviewee. This did not take into account dreams that can be attributed to mere coincidence. Unfortunately, because of World War I and the revolution, this study was never published, and the palm of superiority in the study of dreams was "stolen" from Russia by France and Flammarion.

The unified biofield of humanity

Some modern researchers suggest that our brain can perceive and emit special bio-waves, thus creating a single world superconscious structure that connects all people living on our planet. This structure, according to researchers, has no time, that is, it simultaneously contains the present, the past and the future. Albert Einstein said that "the differences between past, present and future are nothing more than an illusion". But this channel works only for information transfer; external information enters it very rarely, and only in emergency cases. For example, when a person or his relatives are in danger. Perhaps that is why we feel the gaze of others: in primitive times, a glance in the back from a predator or a stranger meant a threat to life.

Based on this theory, some researchers believe that psychiatric diseases begin with a "breakdown" of the mechanism of exchange between the brain and the world information field: a person begins to receive information that he or she cannot correctly evaluate and accepts it as reality. Any doctor has encountered the fact that the mentally ill sometimes clearly predict the future. But with the help of medication "unauthorized access" is quickly blocked and the person comes back to normal. Do you think that's contrary to science? Perhaps. But academician A. N. Krylov believed that "the biggest scientific heresy is that we have already reached the limit of knowledge about the world or are about to reach it".

The pioneer of aircraft construction John William Doonne, who built, among other things, the first British military airplane, was a famous visionary. He paid a lot of attention to his dreams, and in 1927 even published a book "Thoughts on Time", in which he tried to understand the phenomenon of prophetic dreams. Dunn claims to have had his first real dream in 1916. He and before that, he had had prophetic dreams, mostly relating to various worldly and everyday events. But that night he dreamed of an explosion at a munitions factory that killed many people. Dunn tried to appeal to the management of the factory, but he was ridiculed. Two months later, in January 1917, an explosion at a London bomb factory killed seventy workers and injured more than a thousand.

In another dream recorded by Dunn, he saw a newspaper with a report that a volcanic eruption in the Far East had killed four thousand people. Less than a week later, the letter carrier brought him a newspaper with the same headline, but the number of victims was about forty thousand - the scientist was ten times wrong.

Dunn believed that the unified biofield of mankind warns especially sensitive people about coming tragedies, and by taking this into account, it is possible to significantly reduce the number of potential victims.

In 1966, Dr. J. C. Barker, an English psychiatrist from Shrewsbury, became interested in Dunn's research. In England there had just occurred the so-called Aberfan tragedy - a collapse in a coal mine that killed one hundred and forty-four people. Barker published in the newspaper "Evening Standard" a request to respond to those who saw in a dream warning of the coming disaster. He received about a hundred letters with stories of prophetic dreams about Aberfan. Thirty-five letters were corroborated by eyewitness testimony: the dreamers told their family or acquaintances about the dream.

Barker concluded that in the future it would probably be possible to predict disasters by studying people's dreams. However, Barker never explained how people dream, probably because the answer would be far removed from both science and the medicine he practiced. But it is worth recognizing his undoubted research contribution to the study of prophetic dreams.

American professor William Cox was also engaged in similar studies. He managed to prove by means of statistics that trains and airplanes that are involved in disasters are steadily bought fewer tickets than similar flights without accidents. His study was quite extensive: Cox analyzed more than a hundred "catastrophic" flights and more than ten thousand "ordinary". He himself believed that people receive news of impending disaster in their sleep, but this has not been documented.

It is not necessary, we think, to dwell only on dreams: the human brain has many more ways of receiving information, of which official science is not yet suspected. We will talk about it in the next chapters.

The prominent Russian brain researcher, neurophysiologist and academician Natalia Bekhtereva, surprisingly enough, believed in prophetic dreams. Although she said that dreams could not have any connection with the future and therefore dreambooks should not be taken seriously, she always stipulated (partly contradicting herself) that in her life "there were several dreams that turned out to be prophetic. And one of them was incredibly prophetic, down to the details. It was a dream about my mother's death. My mother was alive and well, resting in the south, and shortly before that I received a good letter from her. And in my dream, when I fell asleep in the afternoon, I dreamed that the letter carrier came to me with a telegram informing me that my mother had died. I went to the funeral, met people I had never seen before, greeted them, called them by name - all this in a dream. When I woke up and told my husband my dream, he said: "Do you, a brain specialist, really believe dreams?" To make a long story short, even though I was firmly convinced that I needed to fly to my mom's house, I was talked out of it. Or rather, I let myself be talked out of it. Well, ten days later everything happened exactly as it was in my dream. And to the smallest detail. For example, I had long ago forgotten the word "village council", I had never needed it. But in my dream I was looking for the village council, and in reality I had to look for it - that's the story. It happened to me personally, but I am not the only one. There are many other cases of prophetic dreams and even scientific discoveries in dreams.

There is no way to explain it. It is better not to be wisecracking and say straightforwardly: since it cannot be explained in any of the modern scientific ways, we have to assume that the future is given to us in advance, that it already exists. And we can, at least in a dream, come into contact with a higher Intelligence, or with God - with someone who has knowledge of this future. I would like to wait for more definite formulations, because the progress in the technological direction of brain science is so great that maybe something else will be discovered that will shed light on this problem as well".

Extreme Situations Training

It is not always the case that a nightmare seen in a dream indicates that something is wrong with your health. The brain often uses dreams to simulate extreme circumstances. After seeing, or rather experiencing, an extreme situation in a dream, we can cope with it much easier in reality.

Think of dogs that sleep with their paws outstretched as if they were chasing someone. Having received some information from reality, the brain creates a training situation, replaying future possible events at "idle speed". For example, a person witnessed a street fight, and then in a dream took part in it himself: the brain, based on the experience gained through the senses, teaches him how to act in such a situation.

The famous biologist M. Berg once told that he had a dream in which he was driving a car and a traffic controller waved his baton, showing him that the passage was closed and he should take a perpendicular street. As Berg turned the corner, a heavy truck suddenly came around the corner and he was unable to avoid the collision. The biologist didn't find out how it ended - that's when he woke up. In the morning, he went about his business, and at the same intersection as in the dream, the adjuster ordered him to turn onto a side street. This startled the scientist so much that he did not turn, parked, and saw the truck he had dreamed about come around the corner at high speed.

Dream Workshop

Japan has created a device that can simulate dreams. The device is called "Yumeni Kobo", which means "dream workshop". The principle of operation of the device is simple: before going to sleep, a person looks at the image of what he wants to see in the dream. After that, a story is recorded in the device's memory, which will become the scenario of a future dream.

The machine plays the recorded voice to the person, amplifying the sensations with light and sound signals, as well as special odors. After eight hours, the machine gently wakes up the user, imitating the sunrise with the help of intensifying light (it is known that a sudden awakening can cause sleep to be forgotten).

So far, even by the admission of the developers, the machine has quite a lot of shortcomings: some users complained that in their dreams were not observed the plot sequence, and some even woke up from the sounds produced by the machine. But nevertheless, a significant part of those who have tested the machine's action say that they saw exactly what they wanted to see.

However, many people learn to control their dreams without any machines. Dreams are a treasure trove of knowledge and experience," says Tibetan Lama Tartang Tulku, "but they are often underestimated as a means of learning about reality. In the dream state, our body rests, but we can see and hear, move and even learn. Proper use of sleep is equivalent to doubling our lifespan: instead of a hundred years, we live two hundred.

The Mexican Indians also spoke of dreams in a similar way. The books of ethnographer Carlos Castaneda describe in detail how an Indian shaman taught him how to manage his dreams and even get the necessary information from them.

In fact, everything that is known to our subconscious mind, but is hidden from us, we can learn from dreams. Such practices are now called "conscious dreaming" (from the English expression lucid dreaming). This is a special altered state of consciousness, in which a person realizes that he sees a dream and can control its course. Many books have already been written on this topic, and there are websites dedicated to lucid dreaming on the web. People practicing this technique claim that they experience sensations in dreams that exceed the quality and reality of waking sensations.

If you decide to try the technique of conscious dreaming, before you go to sleep, try to formulate the problem that you want to solve in your dream. Try to visualize all the conditions of the problem, down to the smallest detail, and as you fall asleep, think about the fact that you need to find a solution. Perhaps, while you have little experience in dream management, you will not dream the solution, but it is not excluded that the next day you will suddenly realize how you need to act.

Never be afraid of difficulties and dangers that you encounter in dreams. You have met an enemy - defeat him; you realize that you can not cope alone - call for help from family and friends or think of an almighty assistant. Fell into the abyss - do not be frightened, try to fly to the bottom, but at the same time try to slow down your flight. To do this, first of all, according to experts, you need to realize that you are asleep. To do this there are several simple techniques. You can try to pinch your nose with your fingers, and if this does not prevent you from breathing, it means that you are dreaming and have already started to control it. You can also try to pierce your palm with your finger, try to take off, or look at your hands or in a mirror: your hands or your reflection in the dream will not be stable (as in reality), but shaky, ghostly, constantly changing.

Although the phenomenon of "conscious dreaming" is not yet sufficiently studied by modern medicine and psychology, it is already widely used for psychotherapeutic purposes.

Psychoanalysts believe that many of our problems stem from psychological traumas that were so severe that our consciousness has hidden them deeply and therefore we cannot understand the cause of our current problems. During sleep, however, there is no behavioral response, and so conscious dreams can help us not only to remember traumatizing events, but also to relive them anew, "reprogramming" a favorable outcome. This helps many people to get rid of complexes, fears or phobias.

CHAPTER VI. FORGOTTEN MEMORIES. HOW TO REMEMBER THINGS YOU DON'T EVEN KNOW ABOUT. BRAIN EXERCISE

The famous Soviet psychologist and philosopher S. L. Rubinstein once said: "Without possessing memory, we would be creatures for an hour. Our past would be dead to the future. The present would have irrevocably disappeared into the past. Man would not be able to utilize the knowledge, skills, abilities and experiences of his predecessors. Nor would there be a mental life that unites the consciousness of the individual into one whole, and it would be impossible to carry out the continuous learning that continues throughout our lives and makes us what we essentially are."

The question of how memory is organized has occupied mankind since ancient times. Aristotle, the father of philosophy, and perhaps almost every person who has ever thought about the meaning of life and the frailty of all things, has thought about it. Since the time of the ancient physician Galen, who believed that our impressions are stored in special ventricles of the brain, science has gone far ahead.

Two memory storages

Today, most experts adhere to the theory of Canadian scientist D. Hebb, who in 1949 published a hypothesis about the duality of memory traces and identified two "storages": short-term and long-term. In his opinion, short-term memory represents the first stage in the formation of an engram, or memory trace. It is characterized by the time of information storage from fractions of seconds to tens of minutes and is easily destroyed by such influences as electric shock, head trauma, anesthesia. The amount of information here is limited, and later traces displace earlier ones.

Long-term memory transforms impressions into a stable state. It is not subject to destructive action, and its storage time is not limited. It is believed that the mechanism of this form of memory is sustained neuronal changes.

Thanks to short-term memory, we "live": we remember where we are, who is around us, what actions to take, and so on. Long-term memory allows us to refer to past experiences and the results of our past reflections.

It is believed that a large part of the cerebral cortex is involved in the formation and storage of long-term memory, but since the cortex has been studied very little, this is rather an assumption. There is no specific place to store a particular memory: memory "notches" in the cerebral cortex are widely scattered and also repeatedly duplicated. But it has been observed that memory loss occurs most often in lesions of the frontal and temporal lobes of the brain. It is believed that in the process of learning any knowledge the temporal area establishes a connection with the places where memory traces are stored.

A major Russian neuropathologist S. S. Korsakov at the end of the 19th century proved the direct involvement of the hippocampus in the process of memorization. As a result of his research, he determined that patients with damage to both hippocampi have no personality changes, they quite adequately respond to what is happening, but after a few minutes completely forget everything that recently happened. Also these people can perform mathematical and other operations, but are unable to assimilate new information. They, for example, can not watch movies on TV, because during the commercials manage to completely forget the content of the previous part. But if the movie is played in its entirety, they can, without forgetting the content, to see it through to the end. Therefore, it is believed that the hippocampus provides the transfer of information from short-term memory to long-term memory.

The study of memory processes was foreseen at the beginning of the last century by academician I. P. Pavlov, saying that "the real theory of all nervous phenomena will give us only the study of the physical and chemical process occurring in the nervous tissue, the phases of which will give us a complete explanation of all external manifestations of nervous activity, their sequence and connection".

Already in 1943, less than a decade after Pavlov's death, the Swedish histochemist Hiden suggested that the basis of thinking was nucleic acid metabolism.

To date, most popular theories of memory can be conditionally divided into two groups. The underlying concept of the first group is biochemical theories of memory, according to which information coming from outside the brain is encoded in ribonucleic acid molecules or other macromolecules.

The concept of the second group is that the process of memory consists in the creation of a new structured system, in the formation of new connections between nerve cells. There are many supporters of each of these groups of theories among quite reputable scientists. Only one thing can be said firmly enough: we know very little about the processes of memorization.

Many medical institutes are now working on curing Alzheimer's disease, and new drugs are coming to the market almost every month. Interestingly, clinical trials show positive effects of drugs both developed based on the first group of theories and the second. But, alas, all these remedies are still very weak, and it is too early to even talk about the prospects of defeating memory disorders.

"Alzheimer's" is a disease of civilization

As we know, over time, blood vessels undergo changes: their walls become stiff and lose elasticity. This process occurs in the liver, pancreas and other organs. But the aging of cerebral vessels is of most concern to mankind.

With age, any person becomes absent-minded and the ability to concentrate for long periods of time decreases. However, many experts are cautious about the theory that memory function necessarily weakens as well. Although the aging process can certainly affect memory, much depends on its localization. There are many examples of people with weak and fragile brain vessels whose memory is not affected at all. One can remember at least the founder of microbiology, the author of great discoveries of the late XIX century, Louis Pasteur, who lived 73 years, and until his death he remained creative, although at an early age, due to a vascular lesion, he almost lost the entire hemisphere of the brain.

It is believed that it is not only diseases of the brain, but also of many other organs that lead to a decline in memory function. Emotional factors also have a very negative impact on our thinking functions. It is proven that a person who is constantly preoccupied with something, worried about something, the aging of the brain goes faster. In this case, it is not about thinking and fruitful search for solutions to some problems, but about filling oneself with negative emotions. Interesting is the research of Israeli biologist M. Aronson, who in his work published in 2001 proved that hours of watching TV programs worsen memory and increase the likelihood of dementia.

Old age dementia is referred to by many as "the disease of civilization". Indeed, until the middle of the nineteenth century, the average life expectancy of 80% of our planet's population was about thirty-seven years. Today, a century and a half later, in developed countries it is off the charts for seventy years, that is, almost doubled. Kaari Christensen, head of the research team at the University of Southern Denmark, says that life expectancy in Western countries has been growing linearly since 1840, and there are no signs of slowing down.

Because of this short life expectancy, many diseases that develop after the age of 45-50 were hardly observed or even described in historical medical sources until recent times. Although physicians and philosophers of ancient Greece and Rome do mention senile mental decline (which apparently was not a major problem then), it was first described by the German psychiatrist Alois Alzheimer only in 1906, based on an analysis of observations of fifty-year-old Augusta D.

Over the next five years, eleven more similar descriptions appeared in the medical literature, and the authors of some of them already used the term "Alzheimer's disease."

In 1910, Emil Kraepelin identified Alzheimer's disease as an independent disease, but for most of the twentieth century this diagnosis was made only in relatively young patients, between the ages of 45 and 65. It was not until the 1977 Alzheimer's Disease Conference that it was agreed that the clinical and pathologic manifestations were identical regardless of age.

Today, there is a growing layer of people suffering from this disease without the prospect of recovery. Alzheimer's disease begins imperceptibly, more often after 55-60 years of age, usually with a basic lack of hygiene: a person stops washing, cutting nails, changing underwear. Then more and more time he begins to spend in inactivity, and soon appears and the so-called aphasia - the patient ceases to understand the speech addressed to him, responds inappropriately or in general on another topic. Then the patient stops recognizing others, his movements become inaccurate, hallucinations, epileptic seizures, and unstable paralysis may begin.

As already mentioned, there is no cure for this disease yet. One of the most famous people in the world, former U.S. President Ronald Reagan (a very wealthy man, by the way) in November 1994, informed his fellow citizens that he was suffering from Alzheimer's disease and therefore was retiring from public activities. In 2000, his doctor John Hutton admitted that the health of the former president is steadily deteriorating, and he recognizes only his wife: "For him there is only Nancy. But soon Reagan stopped recognizing her as well. Reagan spent his last years in his own home in Beverly Hills and died in 2004 at the age of 93.

But it's not just the rich and famous who suffer. Today, Alzheimer's disease affects about 27 million people in the world. And the number of those who are afraid of getting it, it seems, is many times greater. Experts advise not to worry, because what is destined to be, anyway, can not be avoided, and in addition, anxiety increases the risk of disease. The most optimal prevention of this disease is the development and training of memory, which sometimes fails even quite healthy people.

The "tip of the tongue" phenomenon

Our forgetfulness surprises even ourselves. We have a word on the tip of our tongue, but we can't remember it. And the word is familiar, it is known to us from the time we were young, we use it at least once a week, well, maybe twice a week... But then it slipped out of our minds, and that's it....

The study of this phenomenon was stimulated by a 41-year-old Italian clerk L. S. As a result of a fall from a horse, he received a contusion on the left side of the skull. After the accident, he easily recognized acquaintances, remembered various information about them, but could not remember their names. Names "flew out" of his memory completely, and even his spouse he was forced to call "wife". It seemed to him that he was about to remember the right name, but it never came back to his memory. This happened in 1988, so the research on this phenomenon is not very old.

The average person is thought to suffer from this kind of forgetfulness, on average, at least once a week. "The 'tip of the tongue' phenomenon," says Benet Schwartz, a psychologist at Florida International University who studies the phenomenon, "is a fundamental side effect of the way our brains are organized." Interestingly, "tip of the tongue" is not just a Russian metaphor; it's found in many languages around the world, including Hindi and Arabic, so the sensation of forgetting a word is the same for everyone.

As it turns out, our memory is a pretty cluttered place, and the storage in it seems to be organized out of hand: names are stored separately from faces, words and their meanings also seem to lie separate from each other.

We can't remember the word on the tip of our tongue, not because we have forgotten it, but because we have lost the "shelf" where it lies.

Contrary to popular belief, our memory is not at all like a classic filing cabinet, but rather like a computer hard disk - information is written wherever there is free space at the moment.

For most of the 20th century, scientists believed that our brain had an index system that clearly tells it where to get the next piece of knowledge it needs at a given moment. This theory was called the "direct access" model, meaning that the conscious brain has direct access to the contents of the unconscious. But now, thanks to the Italian L. S., the theory of data storage in our memory has changed. Today, the theory that the brain does not have direct access to memories, but makes guesses based on other information it can recall at the moment is becoming increasingly popular.

The best evidence for this theory is that it is very easy to induce "tip of the tongue" sensations. It is enough to ask a person a question concluding a vague definition of a not very common term. For example, "compulsory gratuitous payment", or something similar - there are enough such questions in crossword puzzles. A certain part of people are bound to say that the word is rolling around on the tip of their tongue, and they are about to remember it, but the wait may be long. And, by the way, the answer is simple - "tax".

Harvard University psychologist Daniel Schacter, scanning the brains of volunteers, came to the conclusion that during the manifestation of this phenomenon, certain areas of the frontal lobes are activated, including the prefrontal cortex and the anterior part of the cingulate cortex. These areas are thought to be responsible for higher brain functions. During the tip-of-the-tongue state, they are likely engaged in two separate tasks. First, they are busy performing a metacognitive function (in our example, trying to realize the factors blocking the search for a suitable definition). At the same time, when they realize that the necessary word is not found, they start randomly searching the archives of the unconscious, or, on the contrary, they search them in order (for example, from left to right), but without any deeper system. That's why the right word is sometimes recalled quickly, and sometimes only after a few hours. It is a matter of simple luck - whether it is closer or farther away in the depths of the unconscious.

As the frontal lobes age, their size and density decrease, and it becomes harder for them to search the rest of the cortex, which is why the "tip of the tongue" phenomenon happens to you more often as you age. We don't lose our memory, we lose the ability to find something quickly in that memory.

Another phenomenon directly related to the "tip of the tongue" phenomenon is memory blocking. Only here it is not a specific word that is forgotten, but something more extensive. For example, what was the movie you saw last week about. You can't remember, that's all. "I'm jammed," you complain, and you're very close to the truth. Just like when some object jams, so it is here - force won't help, you can only make it worse. The more you worry, the farther away from you the clue. Try to distract yourself from the question you are interested in, drink water, talk to someone on a distracted topic, let your thoughts fly far away, and the desired will come to you by itself.

Something like this can happen, for example, if you are suddenly approached on the street with a request for an interview, or if you have just come for an exam. You have learned everything, you know everything, you read a familiar question in the ticket with pleasure, and suddenly you realize with horror that although you have studied it, but... you have forgotten it! Alas, this is a painfully familiar situation.

Excitement often causes people's voice to change, becoming too high or too quiet. The point is that due to excitement or excitation of the speech center, the tone of the laryngeal muscles and vocal cords increases, or, on the contrary, this tone drops to almost zero. Many people call the voice a mirror of the electroencephalogram, the "correct" characteristics of which are a guarantee of effective brain function. If the voice is "clamped", then, according to the feedback principle, the brain is also "clamped".

You must have noticed that people who keep a sober head in critical situations speak without raising their intonation, which means that they always find the right words and make the right decisions.

The first thing to do if your voice fails is to immediately change to your normal tone and volume. This will immediately relieve the overexcitation of the cerebral cortex as a whole. In short - it is necessary to calm down, and you will remember everything.

The "Remember Everything" diet

Memory is somewhat like a muscle, and like almost any human skill, it can be trained. As with any workout, you should start with proper nutrition. So,

- eat more fruits and vegetables;
- Replace flour products with whole grain products;
- give up egg yolks and butter;
- Try to eat skim milk and low-fat dairy products;
- Reduce or avoid lamb and beef;
- drink more water;
- and most importantly, never forget breakfast.

Waking up in the morning, a healthy person who has slept well feels rested. But his brain has been working all night, albeit in a weakened mode, sorting through the day's impressions. It's foolish to think that it doesn't need a refresher. If you want to strengthen your memory, breakfast should not be a comfort to the stomach, such as a cup of coffee and a sandwich, but help the brain. The ideal breakfast in such a case can be:

- a glass of juice and a bowl of oatmeal;
- a bowl of granola with milk and a glass of juice;
- toasted grain bread, cheese, glass of juice.

If you still can not give up tea or coffee, then try to reduce the amount of sugar consumed: although at first it, like caffeine, activates brain processes, but then they come to a halt, and the brain gets tired much faster than without a "sugar bang".

Lunch and dinner can be as you are used to, but as you nourish your body, don't forget to nourish your brain. It needs phosphorus, which aids in the formation of brain cells. Foods that contain phosphorus include beans, celery, cauliflower, cucumbers, radishes, and walnuts.

Sulfur is necessary for the brain to maintain the cells' ability to oxygenate. It is abundant in cabbage, carrots, garlic, figs, and potatoes.

Zinc boosts mental performance and is found in sprouted wheat and wheat bran.

Calcium is necessary for the processes of hematopoiesis. It is found in apricots, apples, cherries, grapes, beets, cabbage, carrots, green vegetables, almonds, oranges, peaches and strawberries.

Iron ensures the normal composition of the blood, without it many vital processes in the brain tissue are impossible. Iron is abundant in beans, mustard, peas, pineapples, tomatoes, rice and shellfish.

Magnesium will relieve you of insomnia and anxiety. It is found in almonds, lettuce, mint, chicory, olives, peanuts, pumpkin, plums, walnuts, whole wheat grains.

Don't forget about the most important vitamins for the brain from groups E and B. Their main source are salad leaves, sunflower oil, nuts, beans, polished rice, cabbage, spinach, grapefruit, melon, avocado, bananas.

The brain also needs plenty of oxygen - and potatoes, parsley, mint, horseradish, radishes, onions, and tomatoes can provide it.

And don't forget to eat one apple a day!

Memory training: simple exercises

If you want to develop your memory, then, we warn you at once, do not try to memorize anything - mechanical memorization does not lead to anything. Approach the process of memorization in a meaningful way.

- After reading a text, try to understand its main points. You can write them down briefly, or simply underline the key points.

- Rereading the text a second time, pay attention to the nuances, try to connect them to the main idea.

- You can then formulate those questions that the text has not answered for you and try to find them in other sources.

By the way, it is believed that older people learn specialized information better than younger people because their memory and thinking are more structured.

If it is difficult to remember a name or term, use the method of association. For example, if you need to remember that the poet's name is Pushkin, think of cannons and imagine a picture that will help you remember the word.

An unremembered entrance code can also be associated with something. The code "740", for example, with the Jewish melody "Seven-Forty", the code "812" - with the domestic war against Napoleon, and so on. If the number is inconvenient, and it is not tied to a specific date, you can somehow modify it. For example, "5312" can easily be turned into "812" by remembering that the eight forms "5+3". The phone number can also be broken down into memorable digits.

Before any important conversation, even with loved ones, always speak your speech to yourself. Determine the main point you want to convey, without getting distracted by trivialities.

When going to the store or on a trip, take the time to make a list of what you need to buy or take with you. When shopping or simply buying groceries, put aside the calculator and try to do all the calculations verbally: this is good for thinking and prevents your brain from "stagnating".

It's good to start learning some foreign language.

And don't forget to make an effort to keep your blood pressure, insulin and cholesterol levels in check. High blood pressure as well as diabetes and cardiovascular disease have been shown to trigger Alzheimer's disease.

Deja vu - a mental breakdown or memories of past lives?

"It's already happened to me! - We often exclaim, "but when...". And we start going through our memories like a rosary. But, alas, the answer never comes. Scientists call this state "deja vu" (from French deja vu - already seen).

Mention of this state is found in Plato in his theory of anamnesis, i.e. remembrance. He, considering the soul immortal, believed that the process of cognition is nothing but the memory of ideas that the soul contemplated before its union with the body.

This state was first described by French psychologist Emile Boirac (1851-1917) in his book The Psychology of the Future. You cannot predict what will happen in the next moment, but as events occur, you remember them and realize that all of this has happened to you once before. But where and when...? The answer to that question, unfortunately, is not to be found. The feeling of deja vu is sometimes so strong that it is remembered for years or even for a lifetime. It is believed that 97% of people have experienced it at least once in their lives, but it is impossible to cause it artificially, and it occurs quite rarely. Scientists believe that the cause of this state is the interaction of processes in the areas of the brain responsible for memory and perception. The most widespread theory says that when additional neuronal connections occur, information may arrive in the memory area earlier than in the primary analysis apparatus, and the brain, comparing the situation with its copy already in memory, decides that it has already happened.

But recently more and more popular in scientific circles is becoming a theory that deja vu arises from the fact that the brain, for example during sleep, has already "played" a similar situation, programming its owner to a certain behavior, and therefore it seems to us that it has already happened to us.

There is also an opinion that deja vu is the brain's natural defense against the new and unknown: being frightened of something unfamiliar, it creates false memories, trying to convince its owner that there is nothing new in what is happening and similar things have already happened once...

However, already from the presence of such contradictory versions can be understood that scientists are still far from deciphering the phenomenon of deja vu. Many prominent researchers and did not accept any of these versions at all. The famous psychologist, one of the fathers of psychoanalysis Carl Gustav Jung, for example, was convinced that he lived a parallel life in the XVIII century. And that is why all deja vu phenomena were for him only bursts of parallel existences.

Jung once saw a picture of a certain Dr. Stucklberger, who lived in the eighteenth century, and immediately declared that the shoes depicted in the picture had once belonged to him, he remembered them from his visions.

Following Dr. Jung, more and more researchers believe that deja vu is just the soul's memories of past lives in other bodies. The only debate is whether the soul can remember the events themselves, or whether it only remembers the emotions that were triggered by them.

Many researchers believe that there are three perfectly rational ways for adults to remember their past lives: hypnosis, meditation, and conscious dreaming. To begin with, you can use the method of conscious dreaming. Before falling asleep, you should set yourself the goal of seeing your previous lives in a dream and keep it in your mind until you fall asleep. In order not to forget the dream, the awakening should be natural, so do not use an alarm clock. Everything dreamed, even if it seems to you that it can hardly have anything to do with previous lives, should be written down. Perhaps, carefully analyzing these records, you will be able to find some "tails" of your previous lives. Then you will be able to formulate more clearly what you would like to see in your next conscious dream.

People familiar with the basics of meditation may be advised to meditate on their intrauterine development, and then try to get into earlier times.

Alas, in most cases it is not possible to remember events from past lives, but the emotions experienced then are well remembered, which can become the key to getting rid of complexes and fears that haunt you in your present life.

Concluding the topic of deja vu, I would like to say that only you decide what this state is: a failure of consciousness or an echo of previous lives. And there is nothing more interesting than figuring it out for yourself.

CHAPTER VII. EMOTIONS AND FEARS-HOW DO THEY ORIGINATE? CAN THEY BE CONTROLLED?

The word "emotion" (from Latin emoveo - to wave) denotes a process reflecting subjective evaluative attitude of a human being (or animal) to real existing or imaginary situations. Although animals are our younger brothers in reason, we are not talking about them now. Since no method has been invented to analyze the subjective experiences of animals, we, alas, have no one to compare ourselves with. Human emotions are generated by experiences of pleasure or displeasure, fear, anger, physical discomfort, etc., and play the role of orienting subjective signals, i.e. are reduced to the process of internal regulation of activity.

The expression of emotions is largely shaped by society. This can be traced, for example, from various ethnographic descriptions: every nation has its own expression of emotions at every time. Maybe you have paid attention to the fact that the facial expressions of people blind from birth are very poor; the fact is that we learn to express emotions mimetically by looking at others.

First, let's define the terminology. Specialists distinguish emotions from feeling, affect, mood and experience.

Feelings, unlike emotions, are object-based: they arise in relation to someone or something, while emotions arise in relation to the situation as a whole. That is, if you are afraid, it is an emotion, and if you are afraid of a particular person, it is a feeling.

Affects, unlike emotions, are short-lived and always have vivid external manifestations. Emotions, on the other hand, are sometimes stingy in their "effects"; they may not be expressed outwardly at all, but they last much longer than affects. Affects are more about what happens to you, while emotions are about what is going on inside you.

Unlike moods, emotions can change quite quickly and flow more intensely.

Well, experiencing is an exclusively subjective-psychic part of emotional processes that does not include physiological components.

The emotions that arise in a person are not necessarily formed by the situation in which he found himself, they can be the result of memories or representations of experienced or only predicted, or even imaginary situations.

Emotions are a social thing, they are transmitted from person to person, and therefore the emotional experience of a person includes not only individual experiences, but also empathy, which can arise both in the process of communication and in the process of, for example, the perception of works of art.

Take action or relax?

Emotions are divided into stenic and asthenic. Stenic emotions stimulate to active activity (enthusiasm, joy, etc.), while asthenic ones relax or paralyze strength (sadness, grief, longing, etc.).

All emotions have a physiological component. It is very simple: our body has to prepare for what it has seen, and emotion is the best trigger for all its organs. For example, if you are suddenly frightened, adrenaline immediately enters the blood, the work of the sensory organs instantly increases, muscles tense up, and the work of the digestive organs slows down, the frequency of the heart muscle accelerates, the orientation reflex is activated, and so on. You, or rather your organism, thanks to the experienced emotion, is ready for (pardon the tautology, but the essence here is just in the parallelism) the test.

Emotions originate in the following way: impulses from external influences enter the brain in two streams. One stream is directed to the cerebral cortex, the second - to subcortical formations (hypothalamus, etc.). It is established that in the subcortical area there are special nerve structures that are centers of suffering, pleasure, aggression, and calmness. Emotions are directly related to instincts: a person, very angry, can grind his teeth, narrow his eyes, take a threatening pose; he behaves exactly like an animal in a similar situation. Mimicry of higher animals - primates, dogs, cats and others, by the way, in this regard, little different from the human. Another thing is that man is much more socialized, and therefore, unlike animals, he has volitional regulation of emotional reactions.

Also. Emotions are distinguished by valence or tone: they are positive and negative. Alas, a person has several times more negative emotions than positive ones. Such is our life. Researches of scientists proved that regardless *of* social status, income level and other differences, any person has only 40% of emotions - positive, and 60% - negative.

It is clear that getting rid of negative emotions is not only impossible, but also harmful. It has been proven that negative emotions are no less important for our body than positive ones. However, as social beings, we are obliged to regulate our emotions without giving ourselves completely at their disposal.

How to manage emotions

We constantly berate ourselves for having lost emotional control in one situation or another. But, alas, as the French say, this is "humor on the stairs": understanding how we should have acted in a given situation comes only when the acute moment has passed, we go "on the stairs" and begin to berate ourselves for intemperance. Of course, fear, anger or excessive joy are not the best counselors. Strong in hindsight, we promise ourselves that next time we will behave reasonably. But, alas, the next time, our emotions, rearing up like horses, are again out of our control.

The easiest way to recognize emotions as atavism, inherited from prehistoric times, a primitive mechanism of adaptation: fear, which made us run away from danger; rage, which mobilizes muscles; pleasure, in pursuit of which you forget about fatigue... The famous Swiss psychologist E. Klapared believes this way: "The uselessness or even harm of emotion is known to everyone. Let us imagine, for example, a person who must cross the street; if he is afraid of cars, he will lose his cool and run. Sadness, joy, and anger, by weakening attention and common sense, often compel us to take undesirable actions. In short, an individual, once at the mercy of an emotion, 'loses his head'."

The Swiss is definitely right about one thing: if our lives consisted only of crossing busy highways, there would be no need for emotions, but, fortunately, life is much richer. Emotional coldness sometimes interferes in business relationships and is quite unacceptable in personal relationships. Although ten out of ten people would like to reduce the manifestation of emotions to a sufficient minimum, completely abandon them can not. Even the ancient philosopher Marcus Aurelius believed the ideal state of mind impassivity, manifested in the experience of a person exclusively reasonable emotions. And the great Renaissance humanist Erasmus of Rotterdam wrote that there is one and only one way to happiness: the main thing is to know yourself; then do everything not depending on passions, but according to the decision of reason.

Nevertheless, many modern psychologists are against restraining emotions. For example, American psychologist Robert Holt believes that the inability to express anger or rage in a given situation leads to a significant deterioration of health, and if you regularly suppress anger, it can lead to the development of such serious diseases as peptic ulcer, hypertension, migraine and so on. Holt suggested learning to express your anger constructively, remembering that a person seized with anger is actually seeking to "establish, restore or maintain positive relationships with others." The main thing is to try to "express one's feelings sincerely, while maintaining sufficient control over their intensity, which is no greater than necessary to convince others of the truth of one's experiences".

But, alas, anger is precisely that: a loss of control, and we try not to fall into anger precisely because we suspect: we won't be able to keep our cool and channel this emotional outburst in a positive way.

What to do? How to be an emotional person, but not let your emotions spill over the edge?

Self-regulation

Mankind has been trying to control its emotions since ancient times. Indian yogis, for example, note that if a person experiences unpleasant emotions, his breathing becomes shallow or intermittent, and if a person is excited, he takes excessively tense postures. Based on these observations, the ancient Indians created breathing exercises that help to achieve complete serenity. Modern man does not strive for absolute serenity, but yogic practices are widely used in various psychophysical techniques, in particular in autogenic training.

This method of mental control was first proposed by German psychotherapist I. Schultz in 1932. The basics of his methodology are still widely used today in case of increased nervous and emotional tension and to overcome the consequences of stressful situations.

The possibilities of autotraining are great and varied. With the help of autogenic training it is possible to change the temperature of the skin by several degrees, to cause a sharp muscle catalepsy, in which there is a narrowing of consciousness and a feeling of weightlessness of the body, which allows you to freely hold it on weight with the support of two points - heels and the back of the head. It is possible to suggest to oneself that "the hand feels nothing" or "the hand is not mine" and painlessly pierce a skin fold. About half of the people who do autogenic training can induce anesthesia in themselves. Some are able to arbitrarily change the rate of heartbeat, for example, by speeding it up to 140 beats per minute or slowing it down to 45.

However, we are not interested in such extreme experiments on their own body, so let's talk about how to influence mental processes with the help of autogenic training.

With the help of autotraining, you can improve your memory or recall something that seems to be firmly forgotten, such as a poem learned in childhood. Autotraining can be used to enhance reactions, improve the ability to concentrate, eliminate or reduce anxiety and excessive emotional tension. Autotraining is widely used in all special units of the world, as it helps a person learn to master the most complex instrument - his body.

Classroom

You are still new to autotraining, so it is better to choose a room where nothing will distract you. Make sure that there are no drafts, good soundproofing and comfortable air temperature. It is not necessary to darken the room, and if you are distracted by the light from the window, just sit with your back to it. Another important nuance: you should choose a comfortable place for classes, giving a sense of psychological safety. When practicing, you should be sure that no one will disturb you and will not take advantage of your relaxed state to your detriment.

Basic postures for training

Before exercising, you should loosen your tie or belt, unbutton the buttons of your blouse or shirt, remove your watch, rings, glasses. In general, your clothes and accessories should not constrain you or make you uncomfortable.

The most common and the most accessible posture for autotraining is the **"coachman on a drogue"**. The most common and most accessible posture for autotraining is the "coachman on a dolly" pose.

Sit on the seat so that the edge of the chair comes to the crease of your buttocks.

Try to spread your legs wide enough to relax the muscles that contract your thighs.

Place the shins perpendicular to the floor, and if there is still tension in them afterwards, move the feet 3-4 centimeters forward until the tension disappears.

Lower your head so that it hangs down, hunched back.

Rocking back and forth, make sure the posture is stable by balancing the bent back and lowered head.

Place your forearms on your thighs so that your hands do not touch. Try not to lean on the hips, preferably the posture should be stable without it.

Close your eyes.

Breathe calmly, as in a dream, breathing in and out through your nose.

You may not find this pose very comfortable at first, but after a few sessions you will be convinced of its advantages. Please note that symmetry is the main thing in this pose and in the following ones, otherwise you will not create a perfect balance and you will get tired quickly.

Another pose is **"semi-reclining in a high-backed chair"** .

Sit in a chair so that your head is tilted back as you relax. The position of the arms and legs is the same as in the "coachman's pose".

Pose **"lying down with a pillow."**

Lie on your back on a bed or couch, with a small flat pillow under your head.

Place your feet shoulder-width apart and relax your legs; your toes will spread apart.

Bend your arms slightly at the elbows, placing them palms down and trying not to touch your torso with your hands.

This pose is convenient for practicing before going to bed or right after waking up.

The pose **"lying down without a pillow" ("Savasana").**

Lie on your back, preferably on the floor, with something relatively soft under you: a rug or blanket.

Straighten your legs and position yourself so that the distance between your feet is 15-18 cm.

Place your arms at an acute angle to your torso with palms facing up.

Do not rush to immediately conduct autogenic training, first practice to be in the postures themselves for five to ten minutes, and only when you feel that they do not cause you discomfort, go to self-induced.

"Mask of Tranquility."

Do not forget that human facial expressions on the principle of feedback affects his brain, and therefore you need to completely relax the muscles of the face, put on the so-called mask of calmness.

To relax the masticatory muscles, without lowering your head, silently say the "s" sound and let your jaw drop back. For a few minutes, watch the masticatory muscles relax, followed by a wave of relaxation throughout the body.

Then relax your tongue by silently pronouncing the syllable "te".

If you don't have the time or opportunity to engage in full-blown relaxation, it can be replaced with a "mask of calmness", which has quite a powerful effect.

Training

The most common mistake of beginners is not being able to achieve the desired state and, as a result, trying to suggest it to themselves. Alas, this is hardly possible: if you try to suggest it, the last crumbs of relaxation disappear. Don't worry, if you don't succeed so far - try again or simply postpone the session.

At the beginning of the session, you should feel the heaviness of your own body. The creator of autogenic training, Dr. Schultz, offered the following "hints" to help you achieve what you want. Place a fully relaxed hand on a balloon or crumpled paper and see if they crumple under its weight. Or put your relaxed hand on a scale and memorize how the hand deviated. If you have trouble feeling heavy, you can always remember these little hints.

After assuming the chosen posture, concentrate on the real gravity of your right (or, for left-handers, left) arm. Observe its calmness and heaviness until the feeling of heaviness, as you relax, spreads to other parts of your body.

Do not try to spread the feeling of heaviness by willpower, remember, you should be passive and only observe your sensations. Any internal work prevents the occurrence of an autogenic state.

As you gain experience in training, you will feel that at first the feeling of heaviness will be replaced by a feeling of lightness, and then you will stop feeling your own body altogether.

The feeling of heaviness in the arms is always brighter than in the legs or trunk.

After you have felt heaviness, you should feel warmth. It is very simple: immersion in the autogenic state is accompanied by redistribution of blood - its content in large vessels and muscles decreases, while in the capillaries of the skin it increases. If you start to relax correctly and feel heaviness, then warmth will follow.

In the autogenic state, your heart rate decreases. You begin to feel a calm, measured pulsation in the body.

Concentrate on your own breath, feel its calmness. Try to merge with it, continuing passive observation.

Pay attention to your abdomen, it should not move while breathing. Feel the warmth in it.

Then, due to the redistribution of blood in the autogenic state, you may feel that the blood flow to the head decreases. This is accompanied by a cool sensation in the forehead area. This is a very good sensation, as the forehead coolness exercise can relieve fatigue, increase efficiency and even cure headaches.

You can go the very simplified way and say mentally the following formula of autotraining:

"I'm resting. Relaxing. I feel free and easy. I am calm. I'm calm. I am calm. My whole body is relaxed. I feel easy and pleasant. I'm resting.

The muscles of the right arm are relaxed. Shoulder, forearm, hand, fingers are relaxed. The muscles of the left arm are relaxed. Shoulder, forearm, hand, fingers are relaxed. Both arms are relaxed. They lie still and heavy. A pleasant warmth runs down my arms. I am calm.

The muscles of the right leg are relaxed. The muscles of the thigh and shin are relaxed. The muscles of the left leg are relaxed. The thigh and shin muscles are relaxed. Both legs are relaxed. I feel the weight of my legs. I'm calm. I'm calm.

The whole body is relaxed. The back muscles are relaxed. Shoulders relaxed. The abdominal muscles are relaxed. The head is free and relaxed.

My whole body is relaxed. I feel light and at ease. I feel good. I'm relaxed.

I'm rested. I feel refreshed. I feel light throughout my body. I feel invigorated throughout my body. I want to get up and get into action. I open my eyes. I am full of energy and vigor. I'm up!"

But we would still recommend doing without unnecessary words. It may be more difficult at first, but it will bring positive results much faster.

It sounds like you've learned how to relax, and it's time to move on to self-hypnosis.

Autogenic modification

Even before you start practicing, you should be clear about what you want to achieve by practicing. For example, to get rid of sudden outbursts of anger. First, create a positive mental image of yourself as you wish to see yourself. Then formulate, can be on paper, a few formulas for self-hypnosis. These formulas should meet the following requirements:

1. Brevity. The shorter the phrase, the better it will be perceived by the brain. The phrase "Calm", for example, is much better than "I am calm and don't let anyone piss me off".

2. Positivity. You must indoctrinate yourself with positive solutions, not negative ones. That is, the formula "I will no longer give in to anger" is unlikely to work, it is much better to replace it with the formula "I am calm in any situation".

3. individuality. Remember that there are no ready-made formulas and ready-made phrases. Each formula and each phrase is designed for one specific person. Something that may have an absolute effect on you, such as the formula "Enough!", may seem like nonsense to another person. So, when making formulas, choose the ones that will be effective for you.

Many people do not like to use formulas, believing that they can create a picture in their imagination, which affects them better than any words. This method may work for you. Use it if it gives good results.

After you've recited your formulas, it's time to come out of the autogenic state.

Output

Stop reciting formulas and concentrate on the fact that you have given your brain, your body the right attitudes, which will surely bear fruit, and on the fact that you are well rested. Slowly clench your fists, feel the power in them, feel how this power passes into your hands and the whole body. Without unclenching your fists, stretch your hands to your knees, and, together with a deep breath, raise your hands, arch your back, turn your face upwards.

Pause for about 1-2 seconds, then exhale sharply through your mouth, unclench your fists and open your eyes. Class is over.

Negative emotions and laughter

It is unfortunately much more difficult to overcome our negative emotions than our anger. Anger, we know, comes from "inside", while negative emotions seem to come from "outside"; life gives us reasons for them. But, in fact, this is not entirely true.

Physiologists say that a person should not let his brain "get used" to negativity and troubles: one should consider problems or the process of dealing with them as a norm of life.

It is necessary as often as possible to create yourself at least a little joy, which will quickly counterbalance the unpleasant emotions you have experienced.

Everyone knows that there are long-livers in the world, but neither a magic diet that helps to live to a hundred years nor a unique geographical point that promotes longevity has been found. However, research data show that the psychological type of a long-lived person is characterized by such traits as benevolence, friendliness, absence of feelings of irreconcilable rivalry and envy.

There are many ways to regulate emotional states, but one of the most accessible is laughter therapy.

This type of treatment was formalized only a few decades ago, but the healing effects of laughter have been known since ancient times.

Even Hippocrates considered laughter as a therapeutic agent, and Dr. Galli Mathieu, who lived in Paris in the XVII century, believed that in addition to medication, any patient needs laughter, and on the back of the prescription wrote down a couple of his jokes. Soon he became so popular that he stopped taking patients and only sent his patients printed sheets, in the title of which was his name, and under it - different jokes, which, according to Mathieu, had healing properties. It is believed, by the way, that the word "halimaty" came from this "printed publication".

In our time, the first to use laughter for medical purposes was the American journalist Norman Cousins. It happened in the 70s of the last century. The journalist, regularly watching movie comedies, managed to get rid of a severe bone disease, and decided to popularize his method. Today, three main directions are distinguished in laughter therapy:

- "classical laughter therapy". Laughter therapists tell jokes and funny stories in individual and group meetings, while their patients watch humorous movies and read funny books. Scientific studies have confirmed that daily listening to ordinary laughter, recorded on disk or cassette tape, can bring a person out of depression without the use of medication. Today there are several thousand laughter therapists in the world;

- "medical clowning" appeared in the 80s in the United States, and its founder was professional circus performer Michael Christensen. He organized a "clown ambulance" service in Boston, which worked in medical institutions. Today, this service has branches all over the world, and in Israel there are even faculties that teach the specialty of "medical clown";

- The "yoga of laughter" was developed by Bombay-based Dr. Madan Kataria. In his classes, people learn to laugh easily, naturally and as often as possible. The Indian doctor's patients learn belly breathing, practice so-called reflexive laughter (deep, one-push exhalation) and so on.

During laughter, a huge amount of endorphins and neuropeptides are released into our bodies, which leads to a reduction in stress levels, headaches and muscle aches. Oxygen circulation in the blood is activated, antibody levels increase, and immunity is strengthened. Laughter can relieve insomnia, relieve physical, psychological or mental tension. Laughter also alleviates asthma attacks and can even lower blood pressure.

In people who watch a "serious" movie, blood flow drops by 35%, and when watching a comedy it increases by 22%. Neurologist William Fry from Stanford University, says that when laughing a person works about 80 muscle groups, and according to the results of research German laughter Gunter Sickle, one minute of laughter is equivalent to 25 minutes of physical activity. By the way, "laughter centers" are already opening in Europe, a kind of fitness centers, where people do not work out on exercise machines, but laugh.

The results of magnetic resonance scans conducted by Allan Reis of Stanford University showed that laughter activates the same brain centers affected by cocaine in particular.

Dr. Margaret Stuber of the University of Los Angeles suggested that a group of children keep their hands in cold water (about 10 degrees Celsius) for as long as they could. And while under normal conditions the children held their hands for an average of 87 seconds, they held them for 125 seconds while watching a comedy. In this case, heart rate, blood pressure and respiratory rate in children practically did not change, but the level of the "stress hormone" (cortisol) in saliva significantly increased. This means that laughter boosts immunity and helps the body to cope more easily with physical stress.

Interestingly, the ability to cheer is not given to a person from birth; it is one of the first skills we begin to master. In the first two months of life, a person learns to smile, and by the end of the third month, he or she begins to laugh. A child of six laughs 300-400 times a day, but alas, this number decreases with every year we live. An adult, for example, smiles only fifteen times a day.

Growing up, it's not just each individual person who laughs less and less. Unfortunately, the same is happening to humanity as a whole.

Not many studies have been conducted on this topic, but all of them are disappointing: for example, in 1930, the average Frenchman laughed 19 minutes a day, and fifty years later, in 1980, he laughed only 6 minutes.

Today, in the twenty-first century, the results seem to be even worse. Since laughter therapy is one of the few therapies that can be practiced without consulting a doctor, it's time to take your health into your own hands.

How to beat the pain

Although pain is not an emotion, painful sensations cause a strong emotional response. Pain is designed to alert a person to danger when, for example, they grasp the hot handle of a frying pan or step on a thorn. People have been known to be devoid of pain; their nerve endings do not report danger to the brain. Despite knowing that "the handle is hot," these people constantly get severe burns, cuts, and so on. So our bodies are wisely designed, but there are some inconveniences: once we have been notified of danger, the pain might have stopped, but alas....

A person tries to get rid of pain: drinks painkillers or alcohol, smokes, thinks of a lot of ways to forget about the annoying sensations at least for a minute. But this is wrong. You can distract yourself for a minute, but then the pain will return, forcing you to make painful attempts to forget about it again and again.

You should not try to distract yourself from the pain, but rather think about it. Your body has demanded attention, it has given you a sign. Show it that you have heard the "SOS" signals, understand them, and take all necessary measures.

Try to feel the pain fully, to enter into it, to reach the bottom of it, to touch its root. When you do this, the pain itself will not go away, but the emotions associated with it will disappear, and you will be able to stop paying attention to it, to forget about it like a fly buzzing on the window.

How to conquer fear

The ability to control one's fear is probably the most important ability a person can develop. It is not only about the banal fear, which is useful in many circumstances, but also about the ability to control one's body. Often the brain, not wanting to go too deep into emotions, disposes of us with the help of various fears. As we know, our brain is a great reinsurer, and these fears are not good for us.

Tom Muzila, a former American special forces soldier and one of the world's legendary "extreme athletes," recounted how he decided to learn how to walk on glowing coals: "Since childhood I had heard that fire burns the skin. Indeed, many cases are known where a person, trying to walk on a fire, gets terrible burns. But this is only because he is not psychologically prepared for such an ordeal. If your mind and feelings object, it is better not to do it. When I started, I hardened the skin on my soles as much as I could and instilled in myself the firm conviction that there would be no burns, that I could walk on fire."

And Muzila actually managed to walk on the coals, the temperature of which was several hundred degrees, and did not get any burns. Surprisingly, no one has yet been able to explain the phenomenon of walking on fire. It is clear only that with the help of the mind a person can, albeit for a short time, to change the physical characteristics of certain parts of his body. Self-immunity helps to resist not only fire, but also other traumatic influences: mechanical, chemical, etc. We are not yet able to explain this mystery of the brain.

Tom Musila believes that our fears are false certainties, illusions that become psychological realities. When confronted with a phenomenon, a person considers his first impression, which can often be wrong, to be the truth, and begins to believe in this truth, without even trying to overcome this "unconditional fact". Muzila's advice is to perceive reality as an illusion and to believe unconditionally in one's success. It is clear that such a perception requires long training and work on one's emotions.

I realized that what matters is not what happens or may happen, but my attitude to this or that event," says Tom Muzila. - The danger may be very real, the situation may seem hopeless, and everyone around me will be scared, but not me. Whether I am ambushed, hanging on the edge of a precipice, or being attacked by bandits, I have no fear or any other emotion. I only think about how I can best act."

You should not flip through this page just because you are not going to walk on coals yet. The fact is that we are constantly faced with various fears, and they hamper our careers, interfere in personal relationships, and simply poison our lives. The object or situation you are afraid of will not magically disappear from your fear, and you will not be able to think straight and make good decisions under the influence of fear.

As in the case of pain, to stop being afraid, you should not drive away frightening thoughts. You need to figure out why you are afraid, what causes you fear and what scares you most about the situation. Try to break down your fear into its components and understand its cause.

When the cause is discovered, you may be surprised at how ridiculous and insignificant it is.

Many psychologists advise to visualize in detail the situation that scares you, and try, although it is terribly not wanted, mentally relive it. Then again and again. This technique gives good results. First, you will not only get rid of the fear on this particular occasion, and it will not constantly darken your mood. Secondly, if such a situation does arise in your life, you will be ready for it and will endure it with minimal losses.

Erich Fromm in his book "Anatomy of Human Destructiveness" said: "Fear, like pain, is a very unpleasant feeling, and man tries to get rid of it at any cost, There are many ways to overcome fear. For example, medication, sex, sleep, or socializing with other people. But one of the most effective techniques for displacing fears is aggression. If a person finds the strength from a passive state of fear to go on the attack, the agonizing feeling of fear immediately disappears".

This is very good advice: don't be afraid of what hasn't happened yet, try to "deal" with it!

But there are some fears that are difficult to explain even to oneself. Their source is social induction. Fear is a social emotion and can spread like an epidemic, infecting sometimes tens and hundreds of thousands of people. Therefore, always try to understand how much of your fear is your own feelings, and how much you "breathed in with the air". Whether, for example, a meteorite will fall on the Earth or not - is still unknown, but your life, passing here and now, can be spoiled by such fear. Many such fears are created by the media: for example, after Chernobyl there was a fear of radiation, and after terrorist attacks - the fear of being a victim of them. Of course, one should not treat such things carelessly, not paying attention to the bearded man sitting next to you in the subway with some strange things on his belt, but still it is better to die instantly and happily than to suffer from fear for several more years, trying to look under the coat of any neighbor on the bus.

Sometimes mass fears are limited to the limits of one collective or another microsociety. Such fears include, for example, fear of superiors. Someone alone, failing to do their job or feeling inadequate for their position, can easily infect the entire team with this fear. If this happens, first carefully analyze your work and objectively decide whether you have a reason to fear the boss. Even if such a reason does exist, the worst thing that you face in such a situation is dismissal. Unpleasant, of course, but it is not the end of life and not parting with a loved one. And after all, any person is a bit of an animal, and he always feels, maybe even subconsciously, that he is afraid. And the boss immediately has a question: "Why is this subordinate afraid of me? Is he doing something wrong? Is he failing?" They will start to look at you more closely, and in such a situation, mistakes are usually found even where they are not.

If you suddenly feel frightened, the first thing you should do is try to calm down. Make it a rule not to think about the problem or try to solve it as long as you are frightened.

First, restore your breathing. Remember that as soon as a person gets into a stressful situation, their breathing becomes cramped and shallow. There is a lack of oxygen, which leads to even more bodily and mental tension.

Straighten your torso, put one hand on your abdomen just below your navel (if you can't do this because, for example, you are shy of your colleagues, put your hands on your knees or on the table). Take a slow deep breath through your nose, fill your chest with air, feel that your hand (if you put it on your stomach) is raised. Then just as slowly exhale all the air. Feel your arm go down. Repeat this exercise five or six times.

Then try to relax physically: it has long been proven that mental tension gives physical tightness. And according to the feedback method, by relaxing physically, you will calm down psychologically as well. You can relax according to the method of the Swedish doctor Jacobsen, who recommends to first very strongly tense all the muscles of your body, then, staying in this position for about ten seconds, feel and realize this tension, not forgetting to breathe deeply through the nose. Then you should relax sharply and try to feel the accompanying pleasant sensation, the characteristic heaviness in the muscles. This exercise should be repeated several times.

To complete the process of getting rid of fear, you need to close your eyes for a few minutes and imagine yourself in a place where you feel calm and relaxed: on the beach, at your favorite dacha or at home.

CHAPTER VIII. YOUR SEXIEST ORGAN? THE BRAIN!

Marital sex is good for the mind

Regular sex life is the key to health. This is how we can briefly summarize the discovery made by scientists of the University of Hamburg, who after a number of studies have proved that having sex stimulates the work of brain cells. And all the "blame" for the cortisol and adrenaline produced during sex, which increase the intensity of mental activity (the result, note, remains for a day and a half or two days after orgasm) and give a sense of self-confidence.

The main thing - that the partner was proven, constant, and sex - regular (and even better - conjugal). Sexual contact with a charming stranger or a stranger, although it affects the increased production of endorphin (pleasure hormone), but somewhat over the adrenaline level, so that the overall result is rather negative.

Studies by such experts in the field of brain function as German Werner Gamebmel and American Tony Bartlett also confirm that sex is useful for brain activity. For example, T. Bartlett drew the attention of the scientific community to the fact that during sex in the human body begins to produce the hormone prolactin (prostaglandin), responsible for the development of glial cells and nerve fibers of the brain matter (in addition, this hormone perfectly improves mood and counteracts depression). To prevent Alzheimer's and Parkinson's diseases and other brain diseases, the professor strongly advises not to neglect sexual activity, at least twice a week "practicing" sex as a hygienic and preventive procedure. After all, in addition to the beneficial effect on brain activity, regular sex helps to improve immunity (by increasing the level of immunoglobulin in the blood), improves mood and even relieves chronic pain (for example, premenstrual, thanks to the production of the hormone oxytocin).

While remembering the benefits of sexual release, it is important to remember to use protection to protect yourself from diseases that sex does not cure.

Love addicts

Love is a beautiful and spiritualizing feeling. That's what all writers and poets, directors and clipmakers, Hollywood stars and Russian celebrities tell us. And we believe it. But scientists doubt it; well, they are scientists to question everything in the world.

From their point of view, the "divine and empowering feeling" can be considered as an interaction of a person's previous psychological experience, genetic factors and hormonal processes that have merged into a certain matrix that points you to certain features of a potential partner's appearance, smell and behavior that are extremely "attractive" to you. And the consciousness is practically not involved here, the recognition of these "reference points" takes place at the subconscious level.

However, so far even the luminaries of science admit that to create a hundred percent working model of love they can not, but to understand the biochemistry of brain processes that accompany the feeling of love, they managed to some extent.

It turns out that the romantic response to the presence of another person can be "turned on" by certain biochemicals. For example, substances such as:

pheromones

You've probably heard of them at least by ear. The natural odor of a human being is extremely attractive to persons of the opposite sex. It is what we try to exterminate or block with perfume - the smell of sweat or genital secretions - that determines our attractiveness in the eyes of potential sexual partners. This doesn't mean, of course, that you can go weeks without washing and surround yourself with an aura of stench. No, dirty body odor acts repulsively, but if the body smells only of creams, deodorant or perfume, its sexual attractiveness is drastically reduced. Believe me, a clean healthy body, even heated by work or running, smells great.

Recent research into how pheromones make others sympathetic or influence their behavior has shown that there is a chemical androstenol in men's sweat that has a pungent musky odor that excites and attracts women. But this magical property of sexuality lasts only as long as androstenol is not oxidized by oxygen and surface fat, turning into androstenone, the scent of which is perceived rather negatively by most women (except for those who are currently ovulating - for them it continues to be extremely sexy).

Men, in turn, are highly aroused by substances called copulins, which are part of the female vaginal secretion. Clinical trials have been conducted in which men who inhaled the scent of this substance on certain women found them much more sexy and arousing than other participants in the experiment. And the smell of a woman's secretion at the moment when she ovulates even increases the level of testosterone, the hormone of masculinity, in a man's blood.

• **endorphins**

In the period of falling in love and at the beginning of sexual relations, being alone with the beloved especially actively stimulates the production of endorphins - derivatives of brain secretion, acting as the most physiological and natural tranquilizers, causing feelings of security, calmness and peace, and painkillers.

- **Oxytocin**

During a hug, at the moment of pleasurable physical intimacy and especially sexual satisfaction, the brain produces oxytocin, which acts as a powerful anesthetic and euphoria-inducing substance.

I would like to emphasize that the feeling of falling in love, courtship, first meetings and first sexual contacts is accompanied by the production of a whole bunch of substances that stimulate brain cells and cause a state of happiness, excitement, flight, i.e. act as a natural drug. These are dopamine, norepinephrine and phenylethylamine. All these substances, which are produced by the chemical factory of our own body, belong to the category of so-called amphetamines - drugs that, if artificially introduced into the body, act as a stimulant, causing a pleasant, at the same time intoxicating and ecstatic state, euphoria.

The period of courtship and first meetings is perceived by most people as the happiest time of a relationship. This is a time that you want to experience again, minute by minute, to which you want to return again and again ... But the pleasure factory of our body is not able to work hard, love cannot last forever, including for biochemical reasons - over time, the intensity of experiences with the same partner decreases. The body becomes accustomed to receiving a certain dose of certain substances, and their effects gradually smooth out, becoming less pronounced.

Some people take such circumstances as a given - "we are wonderful friends", "passions have cooled down, but the relationship has become more stable". But some people get hooked on the state of falling in love like a drug. Actually, this is what happens - love addicts fall into a strong dependence on the intoxication of love and love euphoria, which they want to experience time after time. As soon as a new relationship begins to lose the charm of novelty and the sharpness of the experience, the "love addict" leaves his or her fling and hurries to get acquainted with a new partner.

Such people, however, find quite logical excuses for their behavior ("I have not yet met the one and only", "all men are bastards" and so on), not suspecting that the whole matter is a chemical dependence on certain brain stimulants.

Struggle of consciousness with subconsciousness 2: 0

The term "subconsciousness" entered our language thanks to Sigmund Freud, the theorist and practitioner of psychoanalysis. In fact, the great role of sexuality, which determines human behavior, was simultaneously noted by another great scientist, Otto Weininger, in his publications. Today, his name says little to the common man, although he was undoubtedly much more talented. No wonder that his fate - unmerciful to true geniuses - was very tragic, and his name - forgotten. When he was only twenty-one years old, Otto published a sensational work, Gender and Character. The public in 1900, the facts and observations set forth in it plunged into a state of deep shock - because the young genius in his work anticipated and formulated all subsequent theories of Dr. Freud. For example, the theory of bisexuality. The book became a bestseller and within two years was reprinted about 20 times! A year later, Weininger was appointed a professor at the University of Vienna, and organizations of his devoted followers and furious admirers began to spring up in Europe like mushrooms after the rain.

And a year later, shortly after his twenty-third birthday, Otto rented a room in the house where another genius - Beethoven - once died, and shot himself at night, putting a bullet in his heart. This pioneer of psychoanalysis believed that sexuality affects relationships between people, including friendship. That is, if there is sexual attraction between you, you will be best friends (or friends), but if there is no attraction, you will remain neutral or start feuding.

"...Friendship between men outside the sexual, even if unconscious, element is impossible," Weininger asserted. - This is already proved by the impossibility of friendship between men who do not like each other in appearance.

Do you feel that sexual arousal offends the sense of friendship? That no one would ever admit that there is a sexual aspect to their friendship? It is in the affirmative answers to these questions that the work of your consciousness is hidden, which will deny even the obvious, just to disagree with the subconscious.

Weininger believed that in one personality both masculine and feminine qualities can be combined quite naturally, because "man and woman are as if two substances distributed in the most diverse ratios in living individuals, and the coefficient of neither of these substances can be equal to zero. It may even be said that in the realm of experience there is neither male nor female. There is only the masculine and the feminine."

Freud only continued the destruction of the moral foundations of the Victorian era, becoming the herald and theorist of the sexual revolution. Previously, sexuality was assigned the role of only an instinct for the continuation of the species, but Freud proved and theorized that the role of the sexual sphere in human life is much more diverse. After all, sexuality, in fact, determines the functioning of most of the mental processes of the human mind.

The revolutionary ideas of "the bottoms want it, the tops can't" led to the sexual revolution. And an era of sexuality was born.....

One of the main tenets of Freud's theory was that human behavior is determined by the subconscious mind (by which was meant hidden sexuality). Culture and social norms suppress a person's natural drives, resulting in a conflict between consciousness and subconsciousness. Such a theory was of great psychotherapeutic value - after all, a person of the Victorian era, brought up in the framework of a complete ban on sexuality, very much wanted to throw off the shroud of mystery from sexual relations. Even a simple frank conversation with the therapist about sexual experiences caused a storm of healing emotions in the patient. What is a psychoanalysis session? The client lies down on the couch, looks at the ceiling and for forty-five minutes three times a week tells the doctor about all the secrets of his sexual experience, starting from the earliest. The main thing is that you are listened to (for a very decent fee, we note), and frankness helps to get rid of mental traumas and heal the wounds of ego.

Having lived through the era of the sexual revolution, psychotherapy began to slowly give up its position - the former revolutionaries began to seem a bit... old-fashioned to the new public. After all, the younger generation could talk about their sexual secrets not only with friends, but even with their parents or, for example, with their grandfather - and, of course, for free. And how could you prove that it was the psychoanalyst who listened to you better than anyone else? Today, just over one percent of psychoanalysts in the United States work with neurosis-like conditions. Which, of course, does not cancel the historical merits of this method, which in the last decade of the last century found its fans in Russia, where the unconscious is still quite relevant. After all, the permanent state of protracted war between the subconscious and consciousness gives rise to various neuroses and complexes in the personality. If it is still possible to live with complexes, then neurosis is already a diagnosis, i.e. a disease. If you have a neurosis - you are not crazy yet, you just suffer from depression or obsessive compulsive states, fears and other unpleasant mental states. You are one in three (the rest somehow cope with inner tension).

But why can't the subconscious mind and the conscious mind live amicably?

The three desires of the subconscious mind

All our problems come from our desires. When they are absent - it is depression, and when they are a lot, but there is no possibility to realize them - it is neurosis. And it wasn't Freud who discovered this. Siddhartha Gautama Buddha told mankind about the harmfulness of desires. Do you remember his story?

Siddhartha Gautama Buddha was a prince, lived happily for thirty years without leaving the palace. Then he decided to see the world. First he met a miserable old beggar begging for alms, then a disfigured cripple suffering from ailments, and then a funeral procession rolled by. And the gentle prince declared, "The world is suffering." And the cause of suffering is desires. If a man had no desires, he would not be upset, sleeping under the starry sky, and would not dream of his house, wife, children, delicious food, new clothes and so on. Besides, there is nothing permanent in this world, everything passes with time, and therefore sooner or later a person can lose money, health, and loved ones. And life itself is also bound to end someday.

With all this in his mind, Siddhartha began to get rid of his desires by trying to become an ascetic. He sat in the lotus position for three years, not moving and eating whatever came to his lips. Already being two steps away from death, Siddhartha realized that he had been a little mistaken: he had to realize that desires are illusory in themselves, and to give them up, one must simply realize the truth, and it is not worth killing the flesh. The prince had an epiphany and became a Buddha.

Giving up desires is the advice the Buddha gave to mankind. It's a great recipe for mental health, no kidding. But another unresolved problem is that most of our desires are not registered by the consciousness, and therefore a person may simply not be aware of them.

But how can it be so, you say. I know perfectly well that I want to become a boss, get two thousand bucks and seduce my colleague (I also want a new car, a laptop and an iPhone, and...). "And I want the same thing, but I also want an apartment in the center...", "And I want (a list of 1000 items)"....

In fact, our deepest desires are only three - to **live, to lead, to reproduce**. It is these three needs, three basic instincts, that determine all of our behavior.

Let's decipher these three desires of our subconscious mind in more detail.

The need to survive. Behind it is fear: fear of hunger, danger, and the unknown, which makes us act - to look for food (to earn money), to flee from a predator (in the most comfortable apartment), to be cautious about the unknown (for example, a new boss or a new friend of the husband).

The desire to become a leader is dictated by great opportunities: first, to survive (the leader gets more and is the first to "enjoy the spoils"), and second, to personally manage the stability of the "pack".

The instinct of procreation or sexual need. The sexual instinct of the animal world is capable of sacrificing life itself (sturgeons going to spawn as if to certain death, death fights between males) in order to continue their species.

But civilized man has perverted himself to such an extent that he has subordinated even the most natural impulses to his immediate needs. He has completely forgotten that natural instincts are not subject to our reason and are incapable of compromise.

People are not beasts, but they cannot give up their animal nature. At the core of our being are the same instincts. Fear of death, lust for power and sexual desire are the three pillars on which the instinct of self-preservation of the individual, the species, the species, is based. This is a simple postulate... which is enough to turn us into neurotics.

Consciousness, burdened with morals, attitudes, principles, protests against fear (it is shameful, humiliating to feel it), power (it is immoral), sex (it is dirty). The subconscious seeks to maximize everything for survival, while the consciousness threatens remorse and punishment, karma and sin. We don't admit to cowardice, even though the situation is obvious. Everyone thinks they are the only one subject to fear, but fear strikes everyone without exception. We think it's shameful to want to be the boss of a firm, but we claim far greater rights over the souls of our loved ones - "for your own good." And sex? That's where the dirt is! Nevertheless, we only help those we find truly attractive. We become altruistic, but only to get approval from "that cute coworker" or "that sexy upstairs neighbor". Think about it - there is a very clear erotic overtone here!

To give yourself to a charming stranger, whose touch gives you goosebumps and makes you dizzy, or to keep your virginity until marriage? Here is a clear conflict of subconsciousness and consciousness, which leads directly to neurosis. Such a choice does not seem to come to the surface, because our true desires (to give ourselves to a man who does not even stutter about marriage) are kept deep inside the psyche by consciousness and sound reasoning, because they seem "indecent". Sometimes, though, such an "inner conflict" comes to the surface.

For example, a man does not feel sexual attraction to his wife, but does not want to part with her for objective reasons - habit, common home, children. And he forbids intimacy on the side because of moral principles (as someone from the great said, "you can not mean you can not" or "the moral pressure of society is the foundation of your moral values"). Everything is clear to the conscious mind, but the unconscious mind protests: prostate pains, depression, hemorrhoids, overweight.

What do you think of this definition: neurosis is a conflict between the conscious and subconscious mind that is resolved by creating a new problem?

For example, a certain man believes he has stomach cancer. The doctors can't find anything, but he continues to worry. He doesn't care that his wife has filed for divorce, that he is having trouble with his boss, he has a more important reason to worry - an imaginary illness in which he is "hiding" from the cruel world. This invented reason, which awakened the fear of death, once overshadows all the "insignificant" problems of life. Now you can not think about how to make up with his wife, and imagine how she will weep at the funeral of her beloved husband along with the repentant boss. "I'm not a loser, I'm sick." The man thought up a "way out", and immediately he felt better. For a while. After all, the problems displaced by such a neurotic symptom, in fact, nowhere disappeared and continue to accumulate, approaching a critical mass. Neurosis can have many variants, but its main sign is the dominance of compulsive fear or behavior over real life.

Headaches, sleep disorders, osteochondrosis, palpitations, weakness - these are typical complaints of those who have suffered on the battlefield of consciousness and subconsciousness. And besides that - depression, fear, attacks of aggression and further down the list. And not the solar eclipse is to blame here, but the conflict between two hypostases of psyche. And here "without 100 grams", i.e. without a psychotherapist, you can't do without it.

Why are schizoids so sexy?

We don't know about neurotics, but schizoids are extremely sexually attractive to the opposite sex - and this has been proven by scientific research. Picasso, Lord Byron, Klaus Kinski, Michael Douglas and hundreds of other creative and successful men - what do they have in common? They all had very tumultuous sex lives. Well, so what, you'll say, it doesn't matter how it turned out. However, psychologists say that the creative energy of actors, artists, poets and other "Masters of Art" acts as a sexual magnet, which largely explains their lifestyle.

Research proves that creativity in humans is directly related to increased sexual activity. This allows us to answer the question that has long plagued geneticists - why is schizophrenia (a hereditary disease) so widespread? After all, the predisposition to schizophrenia is observed in only one in a hundred. And the answer may lie in the fact that the genetic factor responsible for schizophrenia simultaneously makes people artistic natures, which are attractive to the opposite sex.

Professional artists, performers and poets are more likely to be schizoid, according to Drs. Daniel Nettle and Helen Clegg. And on average, they have five to ten times more sexual partners than people in non-creative professions. The brilliance and singularity of talent is like a peacock's tail attracting a female. According to research by Dr. Jeffrey Miller of the University of New Mexico, a person's most impressive abilities are an unsurpassed courtship tool that attracts sexual partners.

Statistical data is also on the side of creators - the average number of sexual partners is directly proportional to creative activity, which brings us to an evolutionary theory that explains why certain qualities of the schizoid personality type persist so strongly in society.

"Creative people are often perceived as more attractive and are the object of everyone's attention," Dr. D. Nettle noted in his paper. - They are often very charismatic and create works that arouse interest in themselves".

CHAPTER IX. MANIPULATION OF CONSCIOUSNESS: HOW IT WORKS

The 20th century was revolutionary in almost all areas of knowledge: people learned to travel by air, the most remote corners were illuminated by electricity, horses were replaced by cars... It is impossible to list everything that has changed in our lives in a hundred years. Serious discoveries were also made in the ability to manipulate people's minds. Only, unlike the above-mentioned achievements, these successes are not so noticeable, and those who use them are not in a hurry to publicize them.

In the nineteenth century, mind manipulators could be counted on the fingers of a few: representatives of some sects, some, very few, employees of special services, mesmerists, or, to put it in modern terms, hypnotists, confidence swindlers. And the majority of these people did not think that they used any technology or verified methods. All, except, perhaps, only mesmerists, went to their goal with the help of intuition and made many mistakes. And the twentieth century created a whole science of manipulation of both individuals and groups of people. Ways of manipulation invented a lot, and we will consider only the most famous. But let's start, after all, with history.

"Brainwashing."

It is believed that Jonathan Edwards, the founder of the Christian revivalist movement, was the first to use brainwashing technology. Most likely, he accidentally invented the technique during a religious "crusade" in Northampton, Massachusetts, in 1735.

Convincing the audience that they were sinners, Edwards accompanied his sermon with vivid scenes of fearfulness, gradually building up the tension. In this way he discovered that he could get his audience to obey him completely. He was able to simulate conditions under which the old tenets of his audience's lives were erased and replaced by new ones.

"You are sinners! - Edwards preached. - You are doomed to eternal torment!". In the end, two of his listeners decided to commit suicide. One was saved, but neighbors of the "new converts" testified that they were so shaken by the sermon that, although they had received "eternal salvation," they could not forgive themselves for their previous sinful lives.

The point is that brainwashing is called brainwashing because all the old beliefs of the listener are erased and new ones must take their place. Edwards switched to a positive program only at the very end of the sermon, and therefore those who did not finish his presentation were completely under the influence of the negative program.

Four years later, in 1739, another Riwayvalist, Charles Finney, used Edwards' work in his sermons in New York City. Edwards' technology, deconstructed and much more thoroughly developed, is still used today. It is used not only in religious cults, but also in personal development trainings, in the United States military, in various business trainings that have now reached us, and many more....

The principle of this technique was formulated by Academician Pavlov, who distinguished three separate and consecutive stages of inhibition (the state preceding hypnosis). The first is the equilibrium phase, during which the brain gives the same reaction to both strong and weak stimuli. The second phase is the paradoxical phase, when the brain begins to react more strongly to weak stimuli than to strong ones.

In the third, superparadoxical phase, reactions and behavioral patterns turn from positive to negative and from negative to positive.

In the first stage, the manipulator influences the emotions of an individual or group of people in an effort to bring the audience to an abnormal, heightened level of anger or fear. The listeners' ability to reason in such a state is reduced and suggestibility is increased. Already in the first cerebral phase it is much easier to master the minds of the listeners, and it is possible to replace the existing mental programs with new ones. In sects, in addition to psychological influence, physiological influence is also used: fasting, regulation of breathing, chanting mantras, reducing the amount of sleep. It is also possible to influence by light or sound effects, unusual smell of incense or even psychotropic substances.

If you want to see the techniques of modern brainwashing in practice, attend a sect meeting or a network marketing seminar. However, this method is hardly suitable for people who are not completely sure of their psychological stability. Both there and there you will see almost the same picture: while people gather, monotonous music is playing, the monotonous rhythm of which is very close to the rhythm of the heart. So already at the preliminary stage, a significant percentage of visitors (and people with above-average suggestibility turn out to be at such meetings) will fall into an altered state of consciousness, the so-called alpha state. A person in this state is twenty-five times more suggestible than a person in a sane, beta state. The same music (or the same rhythm) is usually played at all services, and those who come immediately, without transition, fall into the alpha state, subconsciously evoking in themselves the feelings they have experienced in previous times.

Then the presenter appears and begins the lecture with an excellent "metronome voice": he speaks a little louder than usual, clearly and in a certain rhythm. This is the method hypnotists use when putting a patient into a trance: words are spoken at a rhythm of 45-60 beats per minute. With this rhythm of speech, the maximum hypnotic effect is achieved. The presenter's speech uses vivid but eerie images: he speaks of illness, poverty, unhappy relatives and other negative things and events.

Then, when the audience "matures", the "positive" part of the process begins. The presenter talks about how your life will change if you buy this product (or believe in God), how this choice will favorably affect you: you will become healthy, rich, etc. Then, often, eyewitnesses come out of the audience: "I was very sick, (poor, fat), but as soon as I started consuming and selling this product (or joined a sect), everything changed for the better!

After hearing a few stories like this, the average person begins to believe that he or she will definitely be able to put his or her life in this direction. At that moment, the host's assistants or managers come up to him and offer either to buy what was offered from the stage or, if it is a cult, to talk "more thoroughly.

In the case of network marketing, after a short conversation, the person takes out his wallet, and then at home (that same evening or a few days later, it depends on the degree of suggestibility) he is genuinely surprised: why did I do it, because I have never been able to sell anything! With a sect, everything is a little more complicated: a person has to come several more times, with each new visit becoming more and more entangled in the network carefully arranged by the brainwashers. Over time, he becomes a kind of drug addict who can no longer, or rather is afraid to, live without the trance sensations he experiences at sect meetings.

The three stages of persuasion

Thank God, in ordinary life we are less likely to encounter sectarians than network marketing representatives. Although meetings with the latter also often leave an unpleasant residue. The fact is that after buying a "product" many clients "come to their senses", do not sell the "product" further and do not bring profit to their recruiters. Therefore, "marketers" are constantly in search of new clients.

To recruit supporters, "marketers" need to get them into the first cerebral phase. They usually have little time, so they use three basic principles of selling a "product" that never change.

1. Seclusion. The meeting is held in a place where the participants are cut off from the outside world. It is better if it is a country villa or a boat, but an ordinary hall without windows will do. Participants are asked to turn off their cell phones and are forbidden to leave the room during the session.

2. Constant repetition of the same thing, but in different variations. The facilitator talks over and over again about the importance of not breaking the contract: "You must not break the contract", "Don't break the contract", "The contract must not be broken", "Never break the contract" and so on. Many times during the session, participants promise both themselves and the facilitator that they will honor the agreements they have made.

3. Blunting the ability to reason. After all the "trump cards" have been laid out for you, the presenter's task is to deprive you of the ability to reason, not to give you time to develop skepticism or doubts. To do

this is simple: you are overloaded with new impressions that your brain does not have time to process. This is followed by a little trick: you are convinced of your strength, that you are in control of the situation and thinking straight. The one who is "leading" you asks a simple question, to which you surely know the right answer. The technique is very simple, but it works without fail: the brain overloaded with impressions, having passed this "test", begins to believe that he has everything under control, and that the emerging picture of the world is correct. Then the presenter pushes your brain again, supplying it with a new portion of information.

It is interesting that in many people the subconscious mind does react to such an overload, trying to tell its owner: "Something is wrong here, something wrong is happening". There is no meaningful reaction, but there is irritation, anger, in short, there is negative energy, which the facilitator immediately lets go in the direction he wants, pulling you deeper into his net.

There is only one way out of such a situation - to turn into a "buddha", to let neither positive nor negative emotions take over. And go home as quickly as possible. But alas, very few people are capable of this.

Mysterious hypnosis

Hypnosis is the next stage of influence on a person. This state of consciousness is characterized by a narrowing of its volume and sharp focus on the content of the suggestion, which is associated with a change in the function of individual control and self-consciousness. The person being suggestible is in an artificially induced state of lethargy - drowsiness or pseudo-sleep. Although this is one of the oldest techniques of influence on consciousness, but the term "hypnosis" itself was introduced in 1843 by English physician J. Brad (Bride).

Hypnosis is widely used in medicine to treat overweight, neuroses of all forms, stuttering, depression, fears, lack of appetite, mood swings, feelings of loneliness, panic and anxiety. It is also used to treat psychosomatic diseases such as peptic ulcer disease, hypertension, bronchial asthma, excessive sweating, fever or chills, excessive sleepiness, nausea and dizziness. Hypnosis is used to get rid of various kinds of addictions, such as addiction to alcohol, tobacco, drugs, gambling addiction or even psychological dependence on people, such as a former partner or boss.

B. V. M. Bekhterev, a prominent Russian physician, defined hypnosis as one of the three basic states of the brain, along with sleep and wakefulness. In a person whose brain works in hypnosis mode, the susceptibility to suggestion increases, the power of the word becomes especially significant for him, and the acuteness of real factors is weakened. It is worth saying that not everyone is equally susceptible to the effects of hypnosis, and experts believe that suggestibility is in direct correlation with the development of figurative thinking. The higher a person's intelligence and creative potential, the more hypnable (suggestible) he or she is.

The phenomenon of mass hypnosis is interesting: in large crowds, mutual suggestion - collective suggestibility - increases. For example, at the beginning of the session, the hypnotist asks those present to give a sign if they feel something unusual. Those who are particularly susceptible to suggestion are the first to react, and then a chain reaction begins. Something similar can be seen at meetings of sects or those involved in network marketing.

Despite the fact that hypnosis is widely used in medicine today, its phenomenon, like the phenomenon of brain response, is not fully understood. Scientists, for example, find it difficult to give a clear explanation for the fact that under hypnosis a person can be plunged into both the past and the future.

If the suggestible person is "sent" to the earliest period of his life, his psyche will be a model of a newborn's psyche with signs and reactions characteristic of this age: crying without tears, so-called eye swimming, sucking and grasping reflexes. It is worth "transferring" indoctrinated in kindergarten - and the adult begins to build houses from cubes, crying and laughing like a child. There is a version that while the brain is in a hypnotic state, the left hemisphere, responsible for logical thinking, passes to the "figurative-associative" right hemispheric mechanism of functioning. In this case, the principles of brain functioning that are characteristic of childhood return. In such a state, a person's learning abilities increase and he or she can absorb a huge amount of information, by adult standards.

One can talk about the possibilities of hypnosis for a very long time. Hypnotists immerse patients not only in childhood, but also in old age, and the reactions of a middle-aged person become "senile": hands tremble, speech becomes slurred, and so on. Does this mean that somewhere in our brains we have a ready-made program of how to grow old?

Milton Erickson's Hidden Hypnosis.

But let's talk about the exciting topic of eternal youth a little later, and now let's return to the manipulation of consciousness. In addition to hypnotists with a rhythmic voice, there are those who try to make their work as inconspicuous as possible.

The field of hypnosis, which we are about to discuss, is very common today. It is named after its creator, the famous psychotherapist and hypnotist Milton Erickson (1901-1980). During his long life he managed a lot of things: he founded a school of psychotherapy, treated about forty thousand patients, defended the degrees of Bachelor, Master of Psychology and Doctor of Medicine at the University of Wisconsin, created the "American Society of Clinical Hypnosis" and became its first president, founded and headed the "American Journal of Clinical Hypnosis", wrote more than 140 scientific articles and books on hypnosis.

The methods and techniques created by Erikson are now dominant in American psychology and occupy an important place in world psychological practice.

Milton Erickson is an example of an American dream come true, a man who "made himself." He was born in a small mining town in the western United States, but shortly after his birth, the Erickson family moved to Wisconsin.

When the boy grew up, it turned out that he could not perceive colors and saw the world only in black and white. But this was not enough for nature, and she deprived Milton of the ability to distinguish the height of sounds. And at the age of seventeen Milton fell ill with poliomyelitis and remained disabled for life. But the disease did not frighten him, and perhaps, on the contrary, awakened in him the desire to work and improve.

Completely paralyzed, Erikson spent hours concentrating on the tiny impulses in his leg muscles, first achieving barely perceptible movements and a year later using crutches. He found a sedentary job in a cannery and saved up money for university. And after his freshman year, he embarked on a perilous canoe trip from Lake Madison across the Mississippi to Rock River and back to Madison.

After graduating with a medical degree, he interned and practiced at a psychiatric hospital in Colorado. Erickson soon accepted a position as an associate psychiatrist at the Rhode Island State Hospital. And a few months later, in April 1931, he was accepted on the staff of the Research Service at Worcester State Hospital, where he quickly rose to become the service's chief psychiatrist. Four years later he came to Eloise, Michigan, and was named director of the Center for Psychological Research and Training at Wayne County Hospital. At that time, he also became an associate member of psychiatry at Wayne State College of Medicine, while serving as a professor of clinical psychology at the University of Michigan.

At Eloise, Erickson began using hypnosis to professionally train residents. Erickson then moved to Phoenix where he established a private practice. There he married and had five children.

There is a very interesting example of manipulation (here in the positive sense) that Erickson himself gives in his article "Hypnotherapy in Pediatrics". When Milton's son Robert fell down the stairs, smashed his lip and drove his front upper tooth into his jawbone, the boy screamed loudly in pain and horror. His parents rushed to him: "We didn't even try to lift him up. Instead, as soon as he stopped to gather air for another scream, I quickly said to him, simply, sympathetically and meaningfully, 'It hurts a lot, Robert. Terribly hurt." Undoubtedly, in that moment, my son realized that I knew what I was talking about. He could agree with me and he knew that I completely agreed with him too. So he could now listen to me and trust me, because I had shown him a full understanding of the situation. Then I said to Robert, "And it will still hurt." With this simple statement I expressed his fear, confirmed his view of the situation, showed him my full understanding and my full agreement with him - because at that moment it seemed to him that his suffering and pain would never end. The next step he and I took was to proclaim (as he paused again for breath): "And you really want it to stop hurting." Again, we were in complete agreement, and I was granting and even encouraging his wish. And it was his desire, coming from within and encapsulating his urgent need. Having defined the situation in this way, I could already propose something and be sure of his agreement with me. So I suggested: "Perhaps it will stop hurting after a while, in a minute or two." This suggestion was fully in line with his own needs and desires, and because it contained the words "perhaps it will stop," it did not conflict with his own understanding of the situation. Thus he could accept the idea and respond to it.

Robert knew that he was in pain, that he was suffering; he saw the blood on the floor and on his hands, tasted it in his mouth. And just like any other man in his place, he wanted to see his misfortune as something special and wished for some special comfort. No one wants to have a trifling headache; if his head aches, let it ache so badly that he, the sufferer, is the only one who can bear it. How comforting human pride can be! Robert's attention was therefore directed to two vital aims, expressed in the following simple statements: "How much blood there is on the earth! What wonderful, healthy, good blood it is! Mother, look closely and you will see. I'm sure it is, but I want you to be convinced too." We examined the blood carefully and confirmed that it was indeed healthy and good, but we still needed to do a blood quality study against the white washbasin in the bathroom. Thus the boy, who had already stopped screaming in pain and fear, was washed. When he was brought to the doctor for stitches, he was first of all asked if he could stand what had been done to his sister once before. The stitches were done without anesthesia, because the boy was interested in the outcome of the procedure".

In fact, this story summarizes the whole essence of Erickson's hypnotic method. Ericksonian hypnosis utilizes a person's inherent ability to fall into an involuntary trance, so everyone is susceptible to it, including those who possess it.

The Eriksonian hypnosis technique consists of the following sequence of steps:

- annexation of the facility;
- Behavior;
- trance induction;
- suggestion, manipulation;
- bringing the subject out of trance.

The main thing, of course, is laid at the beginning of communication, and therefore the first stage is very important, the success of further actions depends on it.

Remember that sometimes you meet an acquaintance who is upset about something, and in response to your complaints he just waves his hand: "It's nothing, and it will pass! He is in a good mood, and he just does not want to be "loaded" with your problems. You can understand both him and you - there is no right or wrong in such a situation. Any of us often ignores both cries for help and requests to share the fun.

But there are some people, they are called "the soul of the company", great conversationalists that everyone loves.

The secret to them is simple, it's three letters - NLP.

Neuro-linguistic programming-NLP.

The founders of one of the trends of practical psychology (Neuro-Linguistic Programming, NLP) Grinder and Bendler tried to understand the difference between successful and unsuccessful psychotherapists. It turned out to be very simple: sought-after specialists always started their work by "joining" the client and only then carried out any influence. The unsuccessful ones tried to influence at once - and did not achieve any result.

So, to begin to influence a person, you must first of all join his mood, show him that you consider his problems and joys your own. Remember, the founder of the Christian Church, Apostle Paul, wrote that with the Hellenes he was a Hellene, and with the Jews he was a Jew. So - there is nothing new under the moon.

One of the easiest ways of joining is through body posture. If you want to use this method to feel the inner state of your interlocutor, you should copy his posture exactly. If he is in a closed posture, that is, with crossed arms and legs, take the same pose. But after 1-2 minutes, change your posture, and if your interlocutor also changes his, it means that you have already joined and are "leading" him.

The second method of attachment is done by breathing. This is a more complex technique of influence, but it is more difficult to notice and, therefore, it is more difficult to close from it. You should start breathing in the same way as your interlocutor, and then, gradually, make your breathing deeper and slower. If your interlocutor's breathing also changes, it will indicate that the joining has been successful. But here there is one difficulty, which appears if the interlocutors are of different sexes. Nature has created us different: men have abdominal breathing, and women - thoracic. In this case, it is desirable to use indirect joining, for example, to start moving a finger or swinging a leg in time with the interlocutor's breathing. These movements should be in his field of vision, but do not emphasize them. Make these movements barely noticeable, so that your interlocutor cannot consciously pay attention to them.

Another way of joining is to copy micro-movements. For example, if your interlocutor touches your lips or forehead from time to time, start doing the same thing. Or another option: indicate the beginning of a similar gesture several times and then touch your nose. If your interlocutor repeats the movement after you, the contact is established.

Another way of attaching is psychological; without being a sophisticated expert, it is almost impossible to detect. There is a strict connection between eye movements and the way of thinking a person is currently using. Try right now to remember what color eyes your acquaintance has. First you will have to mentally visualize his face. A person thinks in internal images, and the image you are using at the moment is reflected in your eye movements.

- Moving the eyes up and to the left or right indicates that the person is creating a visual image (as in our example), and a defocused gaze indicates the same.

- The movement of the eyes to the right or left means that the person creates an auditory image for himself or herself, i.e. mentally imagines the sounds of music, the sound of a familiar voice or words spoken by someone.

- If the eyes move to the left and down, it indicates that the person is thinking about what he or she is going to say in the next moment. Such movements do not necessarily indicate that the person you are talking to is lying to you. Of course, it is possible, but most likely, he is just carefully controlling his speech. And this is important for understanding the essence of the conversation.

- Moving the eyes to the right and down means that at this moment the person is creating or remembering bodily images and feelings: for example, warmth, cold, hunger, fatigue, tenderness, or tastes and smells.

Naturally, not every look and not every eye movement means the creation of some internal images. It can also happen that a person just looks at the surrounding objects, but with a little attention, one can begin to distinguish between gazes. When internal images are created, the eyes twitch with a haze, the gaze focuses on one point, pupils dilate, eyelids blink less frequently. Talk to your acquaintances, follow the movements of their eyes, and in a few days you will be able to distinguish between a normal gaze and a gaze that means the creation of images, in which a person falls into a superficial trance.

If you continue your observations, you will notice that each person has a habitual system of images with which he thinks: some people have visual images, others have bodily or, in other words, kinesthetic images, and still others prefer auditory images.

A person's leading system can be deduced from the verbs they use in their speech. If a person is oriented to auditory images, he or she says "listen", "sounds fascinating", "let's discuss", etc. To join such a person, you should use the same verbs.

A person whose thinking is dominated by visual images finds it easiest to explain his position with the words "let's see", "we'll see", "beautiful perspectives" and the like.

With bodily-oriented people, use words like "feel," "touch," "walk," and the like.

A bit of terminology. A person with predominant visual images in thinking is called a visualist, with auditory images - an auditoryist, living with bodily sensations - a kinestheticist.

Trance state

Milton Erickson proved that everyone goes into involuntary trance from time to time, and therefore all people need trance as much as they need sleep. Involuntary trance is a special state of the psyche in which the mind sorts through thoughts and impressions in an attempt to draw any conclusions from them. Most often, the trance state appears during routine, monotonous work. If you need to turn over, for example, fifty pages of this book, after a while you will catch yourself that, while continuing to keep score, you are thinking about something completely extraneous. This is a trance. A long ride in transportation or waiting for something, such as standing in line, also puts a person into a state of natural trance. By the way, it can be quite deep: who among us has not passed his stop?

The trance state is easy to define. Its main signs are:

- posture immobility;
- relaxed muscles;
- slow breathing;
- fixed gaze;
- dilated pupils;

- slowing of blinking and swallowing movements;

- Reduced response to external noise and delayed motor responses.

But rather than waiting for the person to fall into a trance, you can induce this state yourself. Milton Erickson said that you should **say a few statements about what is happening right now that the person will agree with**. These phrases will put the person in a trance, and then you can say the statement you want, which the person will also agree with.

Breaking stereotypes

There are also more sophisticated techniques designed to get your interlocutor to agree with you. The technique of **"negative reaction"** is designed for those who always do not do what they are asked to do. It is difficult to cooperate with such people, but it is very easy to play with them "from the opposite". If you need, for example, for a person of this psychotype to keep silent, say to someone present, or just theatrically, to the side: "Now he will answer us...". This will lead to the fact that the stubborn person, who is set on denial, will remain silent for quite a long time, especially if you from time to time repeat the same head movement that you made when you first said the key phrase.

Another technique, **"breaking stereotypes", is designed for the fact that a person often goes into a trance without getting a habitual reaction to some action.** Another technique, "breaking stereotypes", is designed for the fact that a person often falls into a trance when he or she does not get the usual reaction to some action. For example, when you see a hand reaching out for a handshake, you immediately turn on stereotypical behavior and extend your hand in response. Everyone has developed a stereotype of behavior when the phone rings, when the doorbell rings, etc. Communication in the team and in the family more than half consists of such stereotypes, they are necessary for a person to relieve the mind from routine work.

But if you stop your hand that is already reaching out for a handshake, you will see how confused your interlocutor will be. Stand up and walk out of the office when you hear the phone ring, and your colleagues will be confused.

Here is a common example of a stereotype breaking technique. When saying goodbye, the interlocutor extends his right hand to you, and when you extend your hand to him, he intercepts it with his left hand and says a code phrase that programs you for actions favorable to him. And completes the handshake with the right hand, as usual.

Another way of manipulation is the **"bamboozling"** technique. It is usually used by gypsies and sales agents. The manipulator talks a lot, jumping from one topic to another and constantly demanding confirmation from you: "You see, yes?", "That's right, isn't it?". His speech is quite meaningful, but the phrases are not quite logically constructed, and their meaning is not immediately understood. Your brain does not have time to realize this fast but complex speech, you get lost, and then your interlocutor says a key phrase, with which you most often agree, because it is easier for you to agree than to listen to everything from the beginning. With each new confirmation, you become more and more entangled in the networks set up by the skillful "talker".

There are so many ways of manipulation that it is impossible to list them in one book. At the end of this chapter, it is worth mentioning one more of them, which is somewhat different in technique from those mentioned above. It is called "anchoring".

The manipulator asks you to recall a moment from your life, for example, a great joy or a great fright. As soon as your gaze blurs and you go into a trance, he will immediately set an "anchor": he will make some movement, say some word, or, most likely, just touch a certain place on your body. In the future, in order to cause you fear or joy, it will be enough for the manipulator to make the same movement or touch you in the same way. For example, if you are ready to make an unfavorable decision, he will be able to use an "anchor" on your fright. Although in reality nothing will change, you are likely to postpone your decision, explaining it with a sudden unexplained fear (which you are likely to attribute to intuition or premonition) that "something is wrong here.

CHAPTER 10. PHENOMENAL POSSIBILITIES OF THE BRAIN. HOW TO DEVELOP PSYCHIC ABILITIES IN YOURSELF

The term "extrasensory perception" comes from the Latin words extra, "beyond" and sensus, "feeling". This is the name given to sensations or feelings that we receive with the help of perceptual mechanisms that have not yet been studied. It is clear that their roots, like the roots of ordinary senses, should be sought in our brain.

A number of specialists believe that such superpowers are associated with abnormalities in brain development.

Massive research into human superpowers began in the late 19th century, and between 1880 and 1940, 145 research reports were published, involving some 80,000 subjects.

The opinions of the scientific world are divided, but the authors of 106 scientific studies have confirmed that the findings are clearly superior to random sampling.

Unexplored mechanisms of perception

Most famous were the seven-year experiments on psychic abilities of Dr. J. B. Rhyne, a scientist at Duke University, the results of which he published in 1934 in a monograph entitled "Psychic Perception."

Dr. Rhyne used in his experiments decks of Zener cards with images of five simple symbols - a cross, a circle, a wavy line, a square, and a star. Dr. Rhyne himself described his work with one of the "psychics," theology graduate Hubert W. Pierce, as follows: "The working conditions were as follows: the observer and the subject sat opposite each other at a table with about a dozen decks of Zener cards and a journal. One of the decks would be handed to Pierce to shuffle.

(He felt this created a stronger "rapport"). The deck would then be placed on the table and the observer would remove it. Following this, Pierce would typically take the deck and raise the top card, holding both the deck and the raised card face down; after announcing the card, he would place it on the table still also face down. The observer recorded his announcement. After five or twenty-five announcements - in general we used both conditions equally - the announced cards were turned over and checked against the announcements recorded in the journal. Each card was reviewed and checked personally by the observer, although the subject, at the observer's request, also participated in the checking by putting aside cards that had already been checked. In the course of such a simple task, it is simply impossible to fool an attentive observer in his own laboratory time after time... Another deck of cards was taken for the next round.

Later, there were quite a few complaints about Rhyne's methodology, with some claiming that it could not protect against fraud. Rein responded positively to the criticism and immediately closed any alleged "holes" that were pointed out to him. But it had no effect on the results.

In 1940, Rhyne and another researcher, J. G. Pratt, summarizing the research of the last decades, published "Psychic Perception Sixty Years Later". This work details thirty-five critical hypotheses whose authors attempted to refute the positive results of superpower research.

Critics have suggested errors in the statistical analysis of the results, bias in the selection of published data, inaccuracies in the reporting of experiments, and many other reasons that could lead to incorrect conclusions. But Rhyne and Pratt convincingly refuted all the hypotheses. For example, the method of statistical analysis used by

Rein was studied and approved by many prominent mathematicians specializing in probability theory.

Also, Rhyne and Pratt showed that six of the one hundred forty-five experimental studies under review could not be explained at all by any of the counter-hypotheses put forward by opponents of parapsychology.

A total of 1,850 attempts were made in Ryn's study to guess the target task, i.e., the five types of cards.

According to probability theory, one fifth of the attempts could have been successful due to random coincidences. However, one fifth of 1850 equals 350, but the cards were guessed correctly in 558 cases.

According to experts, the probability of such a coincidence is much lower than one hundred millionth. Thus, we can say that psychics undoubtedly exists. And although it is a subtle, difficult to prove matter, which manifests itself far from always, to deny its existence is at least foolish.

Г. Price, a chemist at the University of Minnesota, quite correctly observed in his 1955 publication in Science that "scientists must choose between accepting the validity of psychic science and refusing to reckon with the evidence obtained by the traditional scientific method." According to Price, the best experiments can only be considered false in two cases: either by assuming that scientists are conscious deceivers or by assuming that they are all insane.

Sooner or later it was necessary to recognize these arguments, and it happened in 1969: the American Academy for the Propagation of Science granted the researchers who had united in the Parapsychological Association affiliate status.

The following curious experiment, an article about the results of which appeared in Nature, was conducted in 1974 at the Stanford Research Institute in Menlo Park, California. Dr. Harold Putow and Professor Russell Targ collaborated for eighteen months with renowned psychic Uri Geller. In one of the experiments, "the subject was virtually separated from the experimenters in a screened room for a week and asked to reproduce thirteen drawings during that period. Geller did not know who performed this or that drawing, who selected the drawings for him and on what principle the selection was made.

The researchers say the target drawing was chosen at random and created only after Geller was acoustically, visually and electrically isolated from them in a room with double metal walls. The drawings were not discussed by the experimenters or shown to Geller.

With the exception of two experiments, drawings were created in rooms adjacent to Geller's shielded room, with distances ranging from four to 470 meters. In the other experiments, drawings were created in rooms adjacent to Geller's inside the shielded room. The drawings to be reproduced by Geller depicted, for example, a horse, a solar system, a bunch of grapes, a tree, an envelope, and so on.

To evaluate the results, all drawings were given to two researchers who had nothing to do with the experiment. They matched the target task data with the response data without error; the probability of matching was here one case in a million..."

In another experiment, one of the researchers vigorously shook a metal box with a dice, and then put it on the table and asked Geller to name the number that fell out. The purity of the experiment was absolute: the box was closed, and the experimenter himself did not know the number that fell out.

Geller, by the way, after a while it seemed to him that the abilities with which he was endowed by nature were not enough and he began to do elementary (partly unscrupulous) tricks, for which he was repeatedly accused of cheating. But the above experiments, although many later doubted their purity, no one could explain from a materialistic point of view. Later researchers confessed that Geller in some cases tried to cheat and with them, but the fact that his talent was not only in the demonstration of tricks, can be considered proven.

Nine experiments on "farsightedness" were also conducted, already with Price's participation. One hundred places, "targets," were selected near the laboratory. A group of researchers were then randomly selected to go to one of these locations. Price, unaware of where they had gone, sat with an experimenter, who also did not know the researchers' route, in a locked room, and described what those who had gone to see the "target" were now seeing. Price's story was recorded on a tape recorder. A total of ten such trips were made. The "travelers" then had to determine to which "journey" each of Price's ten stories belonged. "Travelers," scientists at the Stanford Research Institute, had nothing to do with the experiment. The voting showed that six times out of ten, what Price saw matched what the "travelers" saw.

"While there are inaccuracies in Price's descriptions," Targ and Putow observed, "they are still accurate enough for the judges to identify.

A number of other experiments have been conducted with similar results. Targ and Putov wrote that their research represents "the first step in the search for such models of cause-and-effect relationships that would enable further analytical and theoretical developments in forms familiar to scientific thinking.

Anticipating disasters

Let us turn to a study, the results of which are known, perhaps, to everyone, even to those who have never been interested in psychic science. The American mathematician William Cox summarized the data of many thousands of railway flights and found that trains involved in disasters invariably carried fewer passengers than trains of other flights. Unfavorable flights, as a rule, lacked five to ten percent of passengers (as well as in the already mentioned cases with airplane crashes), although there were exceptions: in the express train "Georgia", in June 1983, traveling from Chicago to Eastern Illinois at the time of the disaster was only six people. And in the train Chicago - Milwaukee at the time of the crash were fifty-five passengers instead of the usual one hundred and fifty. Cox suggested that on a subconscious level many potential passengers were aware of the impending tragedy and therefore refused to travel.

Dr. Rupert Sheldrake put forward the theory of "morphogenetic fields" in the 1980s. He believed that people, animals, natural formations, and even blocks of information release energy that can be picked up by other people. Sheldrake tried to prove his theory in a very unusual way. Having ordered an artist a simple puzzle with a hidden picture, he showed it to volunteers in different European countries and recorded how much time on average it took to guess. The puzzle, along with the clue, was then shown on British television. Then in continental Europe, where British television was not broadcast, the survey was again conducted, and it was found that now the rate of solving the picture was much higher. It is difficult to say how far this experiment is scientifically flawed, but then again, the area it is trying to explore is not exactly correct in the eyes of classical science either. Nevertheless, the experiment achieved its goal, and Sheldrake clearly proved the existence of "morphogenetic fields".

In another experiment, Sheldrake asked the famous Japanese poet Shuntaro Tanikawa to write two poems identical in rhythm and size to a famous Japanese nursery rhyme known to many generations of Japanese children. One poem by Tanikawa was lyrical, while the other, at Sheldrake's request, had no meaning, being merely a set of rhymed words. Both of these poems, along with a nursery rhyme, were read several times to groups of volunteers in England and the United States, who were then asked to repeat one of the poems. None of the subjects knew Japanese, and the sequence in which the poems were read was chosen at random. 62 percent of the experiment participants chose an old nursery rhyme to repeat, although, based on the law of random numbers, it should have been chosen by about 33 percent of the volunteers. This experiment, of course, is also not quite correct from the point of view of classical science, but so far no one has been able to refute its results using proven scientific methods. So Sheldrake's research is, at the very least, something to think about.

Sheldrake's experiment was repeated on a slightly different level by Martin Johansson, a professor at Sweden's Lund University. In an exam, he handed out sealed envelopes to students, saying that they contained copy paper, when in fact they contained the correct answers to half of the exam questions. Surprisingly, the students did best on exactly those questions whose answers were contained in the envelopes.

All these experiments, of course, clash with William Cox's research: it seems that people can pick up from the information field around us not only poems and puzzle solutions, but also information about upcoming disasters.

There are many testimonies about predictions of catastrophes. Let us not recall the mass of prophecies about the fate of the Titanic, but rather analyze the case of the school in the village of Aberfan in South Wales. Many predictions have been made about this tragedy, and they have been thoroughly researched.

On October 21, 1966, the coal dumps of the local mine, which had been accumulating for many decades and weighed about one hundred thousand tons, moved under the influence of prolonged rains and buried the school under them. The disaster killed 144 people, including 116 children.

As often happens after major tragedies, newspaper editorial offices were flooded with letters saying that many people had foreseen the catastrophe. Dr. John Barker, a London psychiatrist, became interested in the phenomenon. He analyzed many hundreds of predictions and was surprised to admit that about sixty of them were genuine, that is, were supported by some evidence. Most often such evidence was the accounts of witnesses who, even before the catastrophe, had heard from the soothsayer about his strange visions.

The mother of one of the dead schoolgirls, ten-year-old Eril May Jones, said her daughter did not want to go to school, saying she dreamt the school was gone as she was "covered by a huge black cloud". The girl added: "When I die, I will have my friends Peter and June by my side." The mother sent the child to school anyway, but, concerned, still told her neighbor about the strange conversation.

A week before the tragedy, Alexander Venn, a resident of South Wales, reported a vague premonition to his wife: "I think something terrible is going to happen near us. In an attempt to get rid of his lingering premonition, he drew a human head covered with black flakes on a piece of paper.

Two days before the disaster, a London resident had a dream in which a small boy was running down a mountain trying to escape a dark cloud that was overtaking him. The dream was so frightening that the lady told her family and friends about it. It is interesting that many other people have had similar dreams, exactly with a mountain and a running little boy.

Dr. Barker, impressed by the many confirmed predictions of disaster, attempted to create some kind of data bank that would prevent future tragedies. He organized the British Bureau of Predictions, but, alas, it received too much information. It turned out to be almost impossible to pick out the true predictions from the total volume of predictions and, moreover, to determine precisely the place where the tragedy could happen.

A similar idea was tried to be realized in the USA, where the Central Office of Prediction was established. The Americans dealt mainly with scientific predictions, but also took into account the most exceptional testimonies of "prophets". But they could not decide how to distinguish true testimonies from false ones.

Developmental anomalies

While some scientists argued about the proper methodology for studying the psi factor, others, accepting the existence of abnormal abilities as fact, attempted to begin their own investigations from that point.

Psychologists from the University of Zurich have conducted a number of studies of people with so-called paranormal abilities. For the experiment they selected those who have the ability to synesthesia (able to "hear color" and "see sound"). The experiments were conducted over a period of ten years. Volunteers scanned the brain while they looked at a special monitor, which stimulated certain centers of the cerebral cortex. It turned out that these people really "hear" color and feel the "taste" of music - the brain areas responsible for taste or hearing, came into activity when the volunteers could not feel any taste or sound.

Scientists have come to the conclusion that these unique abilities are due to anomalies in the development of the nervous system. Synaesthetes have peculiar "crossroads" in their brains, where the paths that transmit neural impulses from the eyes, ears, mouth, and nose intersect. And it happens that the impulses are not sent along the right route.

Of course, this research did not reveal to us the cause of paranormal abilities, nor did it prove their existence, but nevertheless it was a big step forward: not so long ago, synesthetes were considered, at best, fantasists, and most scientists simply did not believe in their abilities. It is possible that soon science will be able to find in our brains the areas responsible for the perception of the external information field.

Parapsychology research in Russia

In Russia, the history of the study of parapsychology began when the fashion for "stoloverchaschim", or spiritualism, - summoning spirits - came from Europe. The genius chemist A. M. Butlerov was the first known scientist interested in this phenomenon.

After moving from Kazan to St. Petersburg in 1869, he became close to his wife's cousin, A. N. Aksakov, a convinced spiritist.

Butlerov wrote that for a long time he could not believe the evidence of his own feelings proving the reality of what he used to consider contrary to common sense. In the end, he had to abandon his doubts and concede only because "one does not argue with facts".

Trying to put the study of spiritualism on scientific rails, Butlerov attracted to his experiments Professor Nikolai Petrovich Wagner, a famous zoologist and writer. Then Butlerov and Aksakov invited the famous medium Bredif to join them. Soon Wagner published an article about his impressions of his séances with Bredif in the "Herald of Europe".

This material caused a protest on the part of D. I. Mendeleev, and he made a proposal to the Physical Society to establish a commission to consider psychic phenomena. This took place in May 1875, and it can be considered that it was in this year in Russia there was a scientific interest in anomalous phenomena. The commission was attended by D. I. Mendeleev, I. I. Borgman, N. P. Bulygin, A. S. Elenev, N. A. Gezehus, N. G. Egorov, S. N. Kovalevsky, and F. F. Petrushevsky.

In total, until November 11, 1875, there were eight meetings of the commission, at which experiments with mediums were conducted. But the experiments did not convince the Russian scientists. The Commission decided that "on the basis of the totality of learned and seen members of the Commission unanimously came to the following conclusion: spiritualistic phenomena occur from unconscious movements or from conscious deception, spiritualistic teaching is superstition," and was eliminated. In March, 1876, its conclusions were published in the press. April 24 and 25, 1876 D. I. Mendeleev gave two public lectures on spiritualism, where he finally debunked it. Many years later, Dmitri Ivanovich wrote to his son Ivan: "Our investigation, no matter how it was scolded, made a decisive impression in society. Since then, spiritualism as a hand removed.

But still Mendeleev, like any great scientist, was not in a hurry to make final conclusions. In one of his works he wrote: "These phenomena should not be ignored, - they should be carefully considered, to know what in them belongs to the category of known physical phenomena, what belongs to the field of human hallucinations, what belongs to the number of shameful deceptions, and, finally, what belongs to the category of now unexplained phenomena that occur according to the unknown laws of nature.

"Unknown laws of nature" continued to occupy the minds of scientists, and in 1885 the outstanding Russian medical psychiatrist V. M. Bekhterev founded Russia's first experimental psychological laboratory. Four years later, in 1889, Julian Ohorovich, professor of philosophy, psychology and physiology at the University of Lvov, made a detailed substantiation of the electrical induction hypothesis of mental suggestion.

In 1898, A. N. Khovrin, a psychiatrist from Tambov, published in a scientific publication the results of his own research on human skin-optical sensitivity and clairvoyance abilities. In 1900, N. V. Krainsky, a psychiatrist, M.D., director of the Koliov Psychiatric Hospital, published the results of his observations on the clairvoyant abilities of female patients suffering from clicuschism, or "demon-possession." In 1902 Ya. N. Zhuk, privat-docent (Kiev) conducted experiments on the transmission of visual sensations. In 1906 Prof. N. P. Myshkin made a report on his discovery of a new type of biological radiation and its ponderomotor (force) action.

The revolution did not prevent these studies either. In 1918, a prominent Russian physician-hypnologist P. P. Podyapolsky spoke at the Society of Neurologists and Psychiatrists at Saratov University with a report "On guessing thoughts.

In 1919, under the leadership of Academician V.M. Bekhterev, a series of telepathic studies on humans and animals began at the Institute for the Study of the Brain and Mental Activity (Petrograd). In 1920, the outstanding animal trainer V. L. Durov joined these studies.

In 1921, in the zoopsychological laboratory of V. L. Durov conducted 1278 experiments of mental suggestion to dogs (696 of them successful, 582 unsuccessful).

In 1923, a book by B. B. Kazhinsky was published in Moscow, "The Transmission of Thoughts. Factors Creating the Possibility of Occurrence in the Nervous System of Electromagnetic Oscillations Radiating Outward". And in 1925, neurologist T. V. Gurstein and Academician V. S. Kulebyakin conducted experiments in Moscow on suggestion at a distance of 55 kilometers. In April of the following year, Gurstein made a report at the meeting of the Society of Psychiatrists and Neurologists in Moscow "On the perception of all kinds of sensations at a distance" on the results of these studies.

In October 1926, the Society of Neurology, Reflexology, Hypnology and Biophysics at the Institute for the Study of the Brain and Mental Activity established a special Experimental Commission on Hypnology and Biophysics (chaired by V. M. Bekhterev) to study mental suggestion. But, alas, the great scientist died in December 1927, and the commission was dissolved.

But two years later, Professor L. L. Vasiliev became interested in the abnormal abilities of the brain. During his scientific trip to Germany and France, he familiarized himself with the work of the International Metapsychological Institute and the Berlin Institute of Parapsychology and established contacts with prominent foreign parapsychologists. In 1932, experiments began at the Leningrad Brain Institute under the direction of Vasiliev to establish the physical basis of telepathy. Similar studies took place in Moscow, in the Laboratory of Biophysics of the USSR Academy of Sciences, led by Professor S. Y. Turlygin.

In 1959, Vasiliev's book "Mysterious Phenomena of the Human Psyche" was published, by then a corresponding member of the USSR Academy of Medical Sciences. In 1962, two more of his books were published: "Experimental Studies of Mental Suggestion" (L., LSU Publishing House) and "Suggestion at a Distance" (M., Political Literature Publishing House). In the same year in the psychological laboratory of the Nizhne-Tagil Pedagogical Institute experiments were conducted with Rosa Kuleshova, who showed the effect of skin-optic vision.

And in 1965, the first scientific publication on color vision phenomena by R. Kuleshova appeared in the academic journal Biophysics. In 1970, a special commission was established to investigate parapsychological phenomena, approved by the order of the Secretary of the CPSU Central Committee P. N. Demichev.

Leonid Leonidovich Vasiliev, alas, died in 1966. After his death, the laboratory was closed and the unique research was interrupted. Vasiliev's work is difficult to overestimate. He tried to separate superstition and mysticism from real science. In the book "Suggestion at a Distance" Vasiliev wrote: "Eighty-five years of experimental research was not enough time for universal recognition of suggestion at a distance as a scientifically established fact. Some authors believe that, since for such a long period of time it did not happen, then, further efforts are not necessary - suggestion at a distance does not exist. Such a conclusion is, of course, unfounded: questions of psychology are resolved very slowly. It took a whole century (from the end of the eighteenth century to the end of the nineteenth century) to convince the scientific world of the real existence of verbal suggestion and hypnotic sleep, phenomena more common and easier to induce than suggestion at a distance.

Cases of spontaneous telepathy often look more like "scary stories" of science fiction writers than the usual "factual material" of a scientist; naturally, it is repulsive, seems frivolous. In fact, cases of spontaneous telepathy are often certified by archival documents and testimonies of living witnesses no worse than historical facts and data of judicial investigation, on the basis of which the guilt in the crime committed is established.

<...> The theoretical and practical significance of the problem of suggestion at a distance depends on the spirit in which the questions noted in the preceding paragraphs will be resolved, especially the question of the nature of the transmitting factor. If, more than hoped, it is resolved in favor of kilometer radio waves (medium and short wavelength waves are excluded by our experiments with the shielding of the inductor), it will be beneficial, perhaps, in practical terms (the possibility of artificial reproduction of similar waves of high intensity), but not very interesting in terms of theoretical (brain radio communication is no more than that).

The final overthrow of the electromagnetic hypothesis of suggestion at a distance can give incomparably more to science. Then, naturally, the question will arise of the production by the most highly organized matter - the matter of the brain - of an as yet unknown factor, which, it must be thought, is energetic in nature. Of its characteristic properties, we can already point out two: propagation over long distances and penetration through any obstacles. Neutrino particle streams and universal gravitation have such properties, but nothing says that these factors are somehow connected with the work of the brain. So we have to look for something else, something new.

It has happened more than once in the history of science that the establishment of new facts, unexplained by what is known, has entailed the covering up of unforeseen aspects of being."

Telekinesis

The most, perhaps, fabulous and incredible of human abilities seems to be telekinesis, or, as it is more often called in recent times, based on the Greek words ψυχή, "psyche" - "soul" and κίνησιζ, "kinesis" - "movement" - psychokinesis, the ability of a person to influence physical objects with the effort of thought.

"Psychokinesis" is a foreign word, and it was invented in 1914 by American author and publisher Henry Holt. But the term "telekinesis" has Russian roots, and it was first used in 1890 by paranormal phenomena researcher A. N. Aksakov. But over time, these two concepts have slightly separated. Under telekinesis today usually mean kinetic influence on objects, under psychokinesis - any mental influence of a person on physical objects, including pyro- and cryokinesis, aero- and hydrokinesis and so on.

Specialists find it difficult to say when the phenomena of psychokinesis were first mentioned, and usually refer to the Bible. But all, however, agree that they have been most widespread in recent times, beginning in the second half of the XIX century, when at spiritistic séances, in the presence of mediums, observed phenomena of levitation and the like.

The first to study psychokinesis was Joseph B. Rine from Duke University in North Carolina. The subject was a man who claimed to be able to arrange thrown dice in the right combinations by the effort of thought. During seven years of experiments 651,261 dice were thrown. The result obtained, according to Ryan's claims, indicated that the probability of such a coincidence tended to one case in ten to the one hundred and fifteenth degree.

Rein suggested that both extra-sensory perception and psychokinesis have nothing to do with the physical activity of the brain and cannot be explained by the known laws of physics, but at the same time, they are extensions of each other. And both can be regulated by the influence of drugs or hypnosis.

Ninel Kulagina

But, of course, the most famous in the world owner of the gift of psychokinesis (if we do not take into account various showmen, to whom there are a number of questions) was a native of Leningrad Ninel Kulagina. In April 1942, sixteen-year-old Ninel went to the front as a volunteer, where she was wounded in the stomach, underwent five operations and was awarded an order and medals. Ninel knew about her abilities long ago, but she decided to publicize such "obscurantism", contrary to the materialist line of the party, only in the thawing sixties and turned to the Leningrad Institute of Metrology.

There, she demonstrated that she could move matches across a table without using her hands, suspend a ping-pong ball in the air, swing a pendulum under a glass cover, and the like.

After several experiments, the delighted scientists took Kulagina to Moscow to continue her research. There, after a number of experiments, it was decided that "the experiments were conducted incorrectly" and their results could not be believed. Kulagina was declared a fraud. It is not known who prevented the Moscow academicians from conducting "correct" experiments. Apparently, the obtained result frightened them so much that they, contrary to all scientific practices, simply decided to "forget" about the phenomenon.

A number of articles appeared in the press accusing Kulagina of fraud, which in Soviet times was tantamount to a court sentence. But a few years later, fate brought Kulagina together with Academician Yuri Borisovich Kobzarev, Hero of Socialist Labor, winner of the Stalin Prize. Having carefully studied her phenomenon, he demonstrated the woman's unusual abilities to several academicians and Professor Yuri Gulyaev, deputy director of the Institute of Radio Engineering and Electronics, USSR Academy of Sciences. The result of this meeting was the Institute's letterhead with a summary: "The phenomenon demonstrated by N. S. Kulagina is of tremendous interest to science. Its study can lead to fundamental discoveries, not inferior in their importance to the theory of relativity and quantum mechanics". It was a kind of a letter of protection.

Kulagina died in 1990, but the controversy around her phenomenon is still raging. For example, Corresponding Member of the

Russian Academy of Sciences A. M. Ivanitsky believes that during the demonstration of moving objects Ninel Kulagina in some cases used thin threads. He is echoed by the famous Soviet illusionist Yuri Gorny: "...she (Kulagina) used strong magnets and thin threads, imperceptible to the observer. Sometimes she did it in a sophisticated way. For example, she asked matches to cover with a glass, and they still moved, changing the direction she set. Thin steel needles were previously driven into the matches, which were influenced by magnets located in her shoes and in her abdomen.

It is clear that the use of magnets and threads has never been proven by anyone, and what is more, it is even more difficult to believe in it than in the amazing phenomenon itself.

Kulagina demonstrated her abilities not only to many Soviet and foreign scientists, but also to a number of Western journalists who had "eaten the dog" on such tricks and could hardly fail to notice them. In addition, much of what Kulagina demonstrated cannot be explained by the simplest fraudulent techniques. For example, the movement of a plastic ball in a closed vessel, or the movement of a glass shot glass under a glass cover. The same matches before Kulagina began to affect them, passed a thorough inspection. It is also difficult to explain how Kulagina "illuminated" photographic paper at a distance or stopped and then started the heart of a test frog. Moreover, the illusionist Amayak Hakobyan, who was invited as an expert, stated that he had no idea how to repeat some of Kulagina's "tricks," such as scattering a laser beam.

Kulagina probably did have unique abilities, but party ideology and the rigidity of scientific thinking prevented a full investigation of the phenomenon. Scientists were too frightened by the results they were able to get from studying Kulagina's abilities. Kulagina affected objects with ultrasound and electric fields. How her body was able to generate these energies, it seems we will never know.

Training superpowers

It has long been known that even the brain of an ordinary person has a huge unused reserve; in our ordinary life we use only a tiny fraction of its abilities. Therefore, each of us can develop superpowers through simple training.

To begin with, learn to listen to yourself. Our subconscious mind is constantly working, digesting the mass of information coming to it, but not displaying it on the "scoreboard". At the first stage of working on yourself, just try to feel what your brain has felt, without trying to interpret the information received - you will learn this later.

There are many ways to try to see what lurks in our subconscious mind. Magicians capable of predicting the future have long used several "mirrors" of the subconscious mind. The most famous, perhaps, such a "mirror" is a crystal ball. However, any other translucent object will do.

Begin to stare intently at such an object, simultaneously directing your gaze through it, trying to fix what is happening inside the object. If you feel that you succeed in such a "specific" gaze, proceed to practice: put a candle behind the ball, turn off the light in the room and take care in advance that you are not distracted by the phone, your relatives or sounds coming from the street. Try to throw out of your head all the things and thoughts that occupied you all day. Monotonously repeat in your mind the question, the answer to which you want to know. If you are simply interested in "looking into the future", then try to fall into a state of light trance, which has already been described in this book. For this purpose you can, for example, concentrate your consciousness on a repeated mantra or simply on your own breathing. The images you will see in the ball are sometimes very clear and sometimes very vague. Only you can decipher the broadcast of your subconsciousness.

Another method Nostradamus is said to have used is fire. The famous soothsayer used to sit by the fireplace and "fiery messages from the Creator" appeared to him. Indians and shamans of the north also use fire for predictions, and Celtic magicians used it to prophesy.

If you don't see anything in the balloon or in the fire, don't get upset. Either your subconscious mind knows how to keep its secrets, or you just need some more practice. But, frankly speaking, everything that you can extract from your subconscious mind in such a complicated "magical" way can be extracted in a much simpler way - with the help of psychoanalysis or even autotraining.

CHAPTER XI. CAN ARTIFICIAL INTELLIGENCE REPLACE THE HUMAN BRAIN?

You may be surprised, but many analysts of the high-tech market believe that scientific developments in the field of artificial intelligence have reached a dead end. Over the last decades, attempts to create thinking machines have ended in failure and disappointment. How can this be, because computers have become many times more powerful and "smarter" than ten years ago? But if in the case of productivity, science is moving at a breakneck pace, then on the way to creating artificial intelligence, the progress is being made in small steps.

And the fact that a machine beats a world champion at chess is not yet a demonstration of artificial intelligence, it is an indicator of the intelligence of the people who have created a perfectly working program. So don't react too enthusiastically to the loud marketing moves of those who love grants and hype in the press, promising that in a couple more years, and we will begin to communicate with machines on an equal footing ....

But what's the catch, is creating a real working artificial intelligence harder than sending a ship to Mars or sending people to the Moon? Perhaps it's no more difficult than figuring out what artificial intelligence really is - thinking machines, Japanese anime with permanently fighting Transformers, "I Am Robot" with Will Smith, and so on. In popular culture you will find many illustrations of the term. But who is to say what artificial intelligence should be in reality, how exactly it should work? For example, the idea of neural networks - is it already artificial intelligence, or just a more advanced model, not fundamentally different from the previous generation of machines?

What methods or tests can be used to determine that a machine has finally learned to think, that it possesses, already possesses, consciousness?

Turing test and bots

In an article entitled "Computing Machinery and Intelligence" ("Computing Machinery and Intelligence"), published in 1950 in the philosophical journal Mind, Alan Turing proposed an original test (which was later called the Turing test) to determine how close artificial thinking came to human thinking.

The schematic description of the test is as follows: "There are three participants - a person, a person and a machine. All three participants do not see each other. Answers are given at equal intervals, in writing. One person evaluates the actions of the other two participants (the logic of their answers), and the computer must answer the questions in such a way that the judge cannot determine whether he is talking to a human or a program. If the judge makes a mistake and evaluates the machine's answers as human answers, the machine is considered to have passed the test - that is, reasonable.

Since about the early 1990s, there have been annual competitions of programs written to pass ("fool") the Turing test. Participants, prizes and results of the tournaments can be found on the Internet (try Googling).

Such programs have become known as bots - they are so popular in LiveJournal that many users often have a couple of such bots on their personal information page, which can superficially communicate on neutral topics and may well pass for pleasant interlocutors.

But we can't talk about the "reasonableness" of bots yet - they are just masterfully written programs that act strictly under conditions set by a human. The machine, strictly speaking, does not understand the questions you ask - it has standard and maximally neutral answers, which the program gives as a reaction to the question sentence and the key words in it.

Talented programmers try to avoid all possible traps. For example, interlocutors often ask the same question twice, so you need to "teach" the program to answer in such cases: "You have already asked this question". If the question is asked for the third time, you should say something like "It's boring" or "You're repeating yourself".

If a person asks: "How many times have you fallen in love?", the machine replies: "That's a philosophical question. I myself would like to know how many times I have fallen in love". And to the question "How many sad people are there in the world?" the machine gives a similar answer: "This is a philosophical question. I myself would like to know how many sad people there are in the world," and so on.

The program merely manipulates the given symbols (as opposed to the human brain, which gives them meaning).

As John Searle, author of the term "artificial intelligence," wrote, some experts believe that "by putting together the right programs with the right inputs and outputs, they will actually create intelligence. But he immediately makes a strikingly simple and beautiful counterargument: "Take, for example, a language you don't understand. For me, that language is Chinese. A text written in Chinese, I perceive as a set of meaningless scribbles. Now suppose I was placed in a room in which baskets full of Chinese characters were placed. Suppose also that I have been given a textbook in English that gives me rules for combining Chinese characters, rules that can be applied knowing only the shape of the characters. For example, the rules might say, "Take this character from basket number one and place it next to this character from basket number two." Let's imagine that the people outside the door of the room who understand Chinese transmit character sets into the room, and that in response I manipulate the characters according to the rules and transmit back other character sets. In this case, the rule book is nothing more than a "computer program". The people who wrote it are the 'programmers' and I play the role of the 'computer'."

"By performing the manipulations prescribed by the rules," Searle ironized, "I will produce characters, unfortunately also incomprehensible to me. In this way, I will pass the Turing test of understanding Chinese. But still, I don't actually understand a word of Chinese. Furthermore, there is no way for me to learn the language in the system at hand, since there is no way for me to learn the meaning of even a single character. Like a computer, I manipulate the characters, but I can't make any sense of them.

The manifestations of artificial intelligence here are as far away as Copperfield's flights are from real levitation.

Microsoft Terrarium

The amusing people from Microsoft offered programmers to take part in the "Terrarium" contest, solving the questions of creating the most viable form of virtual life (how about a contest for the survivability of computer viruses?). Maybe this is the key to the creation of artificial intelligence, which after all will originate in the depths of a super-powerful computer? After all, for computers to be able to "consciously" behave in the real world, their behavioral abilities should not be inferior to those of the most primitive animals. But no, and here is a blunder - the Russians who won the contest have given their most viable virtual animals (herbivores, by the way, not predators) the simplest set of survival rules: "Predator - run away; grass - eat; relatives are running - run with them". So where are the rudiments of artificial intelligence? (Although, in general, man at first was guided only by instincts). It is just that our external world with its incredible variety of possibilities and variants cannot be modeled within the framework of a computer program.

But what computational procedures can be called intelligent at all? What are the criteria of intelligence (setting aside computational abilities)? Could it be a purely biological phenomenon? Again we come back to the original starting point - how can we define the criterion of a machine's intelligence?

Despite the existence of many approaches to understanding the tasks of artificial intelligence and the creation of intelligent information systems, experts emphasize two main paths or approaches.

Top-Down AI, or top-down approach. The development of artificial intelligence goes in the direction of creating knowledge bases and logical inference systems that simulate mental processes: speech, emotions, and thinking.

Bottom-Up AI , a **bottom-up approach,** also called a biological **approach**. In-depth study of neural networks and evolutionary computations that model human intelligent behavior, bringing computer systems closer to human capabilities.

Apocalypse

"Can a machine think? - John Searle, a specialist in artificial intelligence, asked the burning question back in 1980. - Can a machine have conscious thoughts in the same sense that we do? If by machine we mean a physical system capable of performing certain functions (and what else can we mean by machine?), then humans are machines of a special, biological kind, and humans can think, and, therefore, machines are certainly capable of doing so as well. Then, apparently, it is possible to create thinking machines from the most diverse materials - say, from silicon crystals or electronic tubes. If this turns out to be impossible, we certainly don't know it yet."

What if scientists took the wrong path from the beginning, trying to create ONE thinking machine? If the solution lies in another plane - in the integration of already created artificial intelligence systems into a single grandiose thinking system?

Think about it, the achievements of mankind are not stored in the mind of one single person. Human intelligence is the "common characteristic" of society. Hegel wrote: "It is hard for people to believe that reason is real; but in fact nothing is real except reason; it is absolute power." We comprehend language only by communicating with our peers (remember the stories of the real "Mowgli"), we learn by absorbing the experience of thousands and thousands of generations who lived before us (and thousands of generations will follow). And the vast majority of the Earth's population does not invent anything qualitatively new, only preserving (to the best of their ability) the accumulated knowledge.

First there were tools, then - the rudiments of speech, then - mathematics, astronomy, genetics, nuclear physics... The whole of humanity can be imagined as a **huge self-reproducing brain, a** single mind, which makes a qualitative evolutionary leap, self-developing itself and becoming... superhuman.

Just as brain cells form a grandiose structure, so billions of computers around the world, united (and there is already the Internet, a world network linking them together), could potentially create a superbrain and a supercrazum, united under the guidance of a single program that will allow it to feel alive.

But then the question arises, what will be the consequences of the creation of artificial intelligence for humanity? Will it not be the beginning of a global catastrophe?

Visionaries and soothsayers of the 20th century, science fiction writers most often portray general artificial intelligence as an unstoppable force that attempts to overthrow human power. Whether Omnius in 2001: A Space Odyssey, Skynet in The Terminator, or the astounding The Matrix, artificial intelligence, once out of control, begins to dominate the human world.

If the power of the Unified Artificial Intelligence starts to grow avalanche-like, the world we are accustomed to will come to an end.

How could this happen? Suppose a scientific group creates a potentially powerful artificial intelligence. Wouldn't at least a couple of people in that group be tempted to use this "ultimate weapon" to gain power over the world (assuming they know that another Chinese or American group is already close to such a discovery)? To show the capabilities of the artificial intelligence they have created, they demonstrate it with a network - and that's it, the bird has flown away. A powerful artificial intelligence is quite capable of taking control of the Internet - and all the computers connected to it.

And then it will not be difficult for the Unified Artificial Intelligence, for example, to synthesize the image of the U.S. President (voice, logic, personal qualities - even the image - on numerous TVs and monitors) to control the most influential power in the world. Or, using its superintelligence, to create new biological species, more convenient than humans for its incomprehensible human designs.

And the elementary seizure of electronic control systems of nuclear weapons or all technical devices on Earth? We have given here only three ways to seize power - but the mind, infinitely superior to the human mind, can come up with billions of different ways to achieve its goals. And we will not be able to predict and, moreover, to correct these goals.

Would you say that all this is fantasy? But do you think the creators of the first computers could have imagined that in just a couple of decades the spread of self-copying programs that disable millions of computers or steal money would become a global problem? If you told them about a trivial Trojan horse, they would consider you a crazy dreamer.

Of course, the original program capable of generating artificial intelligence will be given certain criteria for the safety of mankind. But a tiny mistake, an insignificant failure, can creep into them. For example, computer scientists working for the American military industry created a wonderful autopilot program for fighter planes, but they did not take into account just one moment. And when the pilots crossed the equator on autopilot, the airplane flipped over and flew upside down.

Many may recall Asimov's Three Laws of Robotics, which define human safety as the main commandment for a robot - but if artificial intelligence decides that humans, for their own safety, should be isolated from each other and locked in separate cells with full utilities, television and daily injections of tranquilizers or morphine? In the perception of a Unified Artificial Intelligence, the good of humanity may look very strange. It turns out that it is impossible to potentially create a "safe" intelligence of such power due to the lack of adequate human control and management. And we have no idea about the moral side of this new intelligence (and human behavioral norms can be skillfully imitated - remember, for example, Cecatillo).

What if there are several AI-empowered Minds that refuse to live together and start fighting for power or survival? Will the human mind be needed in such a world or, at best, will humanity be an appendage of a new computer civilization?

Followers of the world's major religions believe that the brain participates in the thinking process (reacting to pain, forming the simplest skills, including instincts), but still plays the role of a transmitter of information from our soul, a particle of immortal and timeless spirit. And who knows, if electronic super-intelligence, having appeared in our world, will possess superconsciousness, maybe God or the Universe will endow it with an immortal soul? Perhaps, in this case, mankind will have another chance to use the gray matter of the brain for its own good instead of self-destruction.

CHAPTER XII. CONCLUDING. THE PHENOMENON OF HAPPINESS. DOES THE FEELING OF HAPPINESS DEPEND ON THE WORK OF OUR BRAIN?

"What is happiness - everyone understood it in his own way," wrote Arkady Gaidar in the story "Chuk and Huck". And immediately tried to decipher the concept: "But all together people knew and understood that one should live honestly, work hard and love and protect this huge happy land, which is called the Soviet country.

Today, many people are unlikely to agree with this definition. Therefore, let us try to characterize the phenomenon of happiness more universally.

For example, like this: happiness is a feeling of fullness of being, joy and satisfaction with life.

The rich cry too

A happy person may lack something material, but the feeling of happiness allows him to easily accept all the problems of life. The feeling of happiness can be very short, or it can be very long in time. In general, the phenomenon of happiness is full of paradoxes: it is known, for example, that among the elderly people there are much more happy people than among young people who have enough strength and health.

Let us emphasize once again that scientific research has proven that the feeling of happiness does not depend on material wealth: very happy people can be found among both millionaires and poor people. At a recent symposium in Berlin, dedicated just to the question of happiness, psychologists, for example, came to the conclusion that this feeling can cause only love and friendship. To feel happy, it is not enough to have a lot of money, a successful career or, for example, a nice house.

Biochemical causes of happiness

There are many different reasons for happiness, and it is unlikely that anyone will ever be able to make a more or less complete list of them. But the biochemical reasons that make people feel happy are well known: feelings of well-being and happiness depend directly on the level of serotonin and dopamine in the brain.

Serotonin gives a positive emotional outlook, while dopamine is part of the overall reward system and is produced during meals and sexual intercourse.

The synthesis of serotonin, by the way, depends on the time of day, and therefore the tendency to depression may increase when the daylight hours are shortened. If you live in a place where the day is too short in winter, you can only sympathize. It is safe to say that time of year and geographic latitude are important factors in happiness.

The feeling of happiness is not available to all people. There is such a notion as anhedonia - a state of mind in which a person not only does not get pleasure from anything, but even has no desire to get it. Sometimes cases of anhedonia are so serious that doctors have to resort to medication. But, alas, even strong drugs do not always help.

The phenomenon of happiness is still too little studied by science, although the twentieth century has greatly advanced this research. In 1904, it was hypothesized that signals in the brain are transmitted by chemical substances. A little later, two transmitter substances (mediators) - adrenaline and acetylcholine - were discovered. And then research in this area was forgotten for a long time. Scientists returned to this topic only in the 1970s. In 1976, several more mediators were discovered: Dr. Hughes isolated the substance enkephalin (translated from Greek, this name means "from the brain"), and Professor Cho Hao Lee discovered endorphin, that is, internal morphine, which turned out to be fifty times stronger than the famous drug. The study of endorphins - endogenous mediators formed in the pituitary gland and activating opiate receptors, gives a key not only to understanding the many processes occurring in our body, but also to deciphering the principles of the brain itself. In addition to the fact that endorphins regulate metabolism and homeostasis (constancy of the internal environment), mechanisms of memory, sleep, learning, etc., they also give a person a feeling of joy. Very soon endorphins were called "happy hormones".

It is these "happiness hormones" and the chemical reaction they produce that allow us to derive satisfaction from a job well done, a personal relationship, or simply good weather. Cho Hao Lee said this about endorphins: "I feel that these molecules make the difference between happy and unhappy people".

Scientists believe that it is the lack of endorphins in the body caused by the poor functioning of the pituitary gland that leads people to alcoholism and drug addiction. In an attempt to achieve the necessary balance, such people resort to artificial stimulation. But, alas, this is a dead-end path: although opiate hunger is quenched for a while, the brain, noticing the overabundance of "happy hormones", begins to produce them less and less, which can later lead even to atrophy of the pituitary gland. Subsequently, a person who has replaced natural stimulation with artificial stimulation can no longer stop.

For many years chemists have been trying to synthesize endorphin in its natural form. But so far all these attempts have been fruitless.

Moreover, some experts suggest that synthesized endorphin will not be able to solve this problem: it, like ordinary drugs, will be accompanied by addiction and dependence. There is also a version that extraneous chemical substances will not be able to achieve the same effect that a person's own hormones have on the human body: as a similar example, it is enough to recall the rejection of transplanted organs.

Practice joy!

But do not think that if the pituitary gland produces insufficient amounts of endomorphins, a person is doomed to a dull life without joy and happiness. The production of endorphins can be enhanced or "trained".

Everyone knows that doctors recommend that gloomy and depressed people should exercise more, eat good food, eat chocolate and simply do things that give them pleasure. But that's not all: a person experiencing a deficiency of "happy hormones" first of all needs to develop the ability to be happy, the ability to enjoy life. Without this skill, any sports and chocolate will be useless.

By the way, the ability of a person to make himself happy with the help of certain trainings is scientifically proven. A group of researchers from Madison University in the American state of Wisconsin scanned the brains of people who had been practicing Buddhism for several years. It turned out that the left side of the brain - the "center of happiness" in them is much more active than in ordinary people. This explains their calmness and contentment: "At this point, we can assume with some confidence that phlegmatic Buddhists, who can be seen, for example, in places such as Dharamsala (the residence of the Dalai Lama), are actually happy". Interestingly, the positive state of the brain centers in the Buddhists studied was observed all the time, not just during meditation, hence it can be assumed that lifestyle somehow influences brain function. The parts of the brain that control fear and excitement showed much less activity in Buddhists than in other people. This discovery has allowed scientists to begin developing special meditation techniques to treat depressive illnesses.

The famous French writer Andre Morua said that it is not the facts of our lives that make us unhappy, but our attitude to them. Laboratory rats, having once pressed a pedal that stimulates with the help of embedded electrodes their "unhappy" brain zone, subsequently avoid it. It would be good if people would treat their unhappiness in the same way: too often we keep pressing the "pedal", realizing that this will only lead to negative emotions.

It has long been known that negative emotions not only change the functioning of all systems in the human body, but also change the composition of the blood. Canadian physiologist Hans Sellier gave the following definition of stress: "Stress is anything that leads to rapid aging of the organism or causes disease".

It is inherent in our physiology that we react violently to danger. But as human civilization develops, people's lives become safer and safer, but the body's ability to react remains the same. And often the brain begins to invent dangers to keep this property of the body in order. Alas, such constant combat readiness is not the best for our health. So the best recipe for health is to convince your brain that imaginary dangers are much less serious than it thinks, and, without putting off for tomorrow, start practicing the ability to enjoy life, the ability to be happy.

Conclusion

Every day our brain is attacked by more and more intrusive thoughts, they exhaust us, prevent us from concentrating on solving urgent problems and lead to apathy or depression. "I won't succeed, I'm a failure." "How am I going to tell my mom about this?". "No one in the world wants me." "I can't live without him." "I don't want this life, I'd rather die." "Everyone despises me, everyone laughs at me," etc.....

Have you ever thought that such thoughts make you suffer much more than something that happened (or will happen) in reality? Sometimes, in case of mental pathology (for example, schizophrenia), only pharmaceuticals can get rid of depressive thoughts. But if your illness chart is pristine, try to overcome them with your own efforts.

Two conditions can provoke intrusive thoughts - inactivity and intellectual passivity or, on the contrary, continuous working time, stress, lack of sleep and chronic fatigue.

If you work from home or part-time, if you're caring for a toddler or expecting a baby, and you have tons of free time during which bad thoughts pop into your head, you just need something to keep you busy as soon as possible.

Write a strict schedule for the day, make a list of tasks to be accomplished (including "make sure to learn something new", do your favorite activity, listen to an educational audio book while doing homework, and so on). Reorient your brain to solve specific tasks, direct your thinking in the working direction.

If you are "bogged down at work", you should urgently change the environment by going away for two or three days closer to nature, or devote this time to physical work: fix the roof at the dacha or dig up your mother-in-law's vegetable garden. Believe me, after ten dug up hectares, the negative thoughts that have been tormenting you for the last month will fall away like a clod of dried earth.

Has the heat of passion subsided, but sad thoughts still return? Use the advice we have already given - replace the negative with a positive. Write down the main essence of obsessive thoughts on a piece of paper, trying to fit everything into one sentence. On the back of the piece of paper, formulate a thought with which you will replace the first, negative statement. For example: "The doctors didn't find anything, but I think I have a very dangerous disease, maybe even cancer." In this case, one positive statement is not enough, you need to undergo additional examinations - adults and people responsible for their health should understand this. But positive thoughts can transform everything around you - even affecting your health. So, you write: "I have excellent tests. The doctors said that suspicions of (stones, tumors) can be ruled out with one hundred percent certainty. My health is getting better every day. I feel better and better. I enjoy my healthy and strong body." (Important: when formulating positive attitudes, avoid the particle "not"). Write it down on a piece of paper (and carry it with you at all times), dictate it into a tape recorder, write it on a cardboard card, or type it into your computer's memory instead of your screen saver. And as soon as you catch yourself that negative thoughts have penetrated your brain - immediately take out a sheet of paper with the text and loudly, with expression read it aloud (well, or repeat it to yourself, so as not to embarrass your colleagues). Eventually, you will learn this text by heart and repeat it automatically; that's when the negative thoughts won't be able to break through the armor of your positivity.

And please do not think that bad thoughts are something unimportant. As priest Ignatius Bryanchaninov said, "from accepted correct thoughts all good things are born, from accepted false thoughts all evil things are born". Thought gives birth to emotions, actions, and ultimately to destiny.

Often with our "second mind" we realize that our thoughts are hypertrophied and illogical, and sometimes simply absurd, but we cannot "kick" them out of our consciousness! Orthodox priests sometimes speak about being possessed by demons (not those with hooves and pitchforks, but energy demons), who inspire evil and unwholesome thoughts.

From a psychiatric point of view, a person only needs to abstract from what they think are their own thoughts to banish them from their mind. And the place to start is by recognizing that you DO have intrusive thoughts (do you think it's easy for an alcoholic to take that step?), and that you want to get rid of what is poisoning your body and your life.

Take responsibility for your actions (they are not your thoughts, but your own - even if under the influence of depression or negativity coming from obsessive worries). Tell yourself - who is the master in the house (your brain)? Imagine that your brain is a big apartment. Here is the room "work", here are "friends", here are "parents", "children", the pantry "entertainment", the balcony "goodies" and so on. But all the rooms are currently occupied by an intrusive unpleasant stranger - a negative thought. Grab a gun or stun gun, mop, broom or rag (depending on your bloodthirstiness) and chase the uninvited guest (thought) out the door. Open the door (in the area of the frontal chakra) and - away! And then close the door firmly and do not let the uninvited guest back in, because you are the master in the house.

Another important point: refuse the hidden benefit that obsessive compulsive disorder carries. "How can it be so! - the indignant reader may exclaim, "I'm suffering a lot, what benefits can there be, what are you talking about?" About the fact that the flow of obsessive obsessive thoughts gives some people an excuse for their own inaction, passivity and laziness ("I'm suffering, I have obsessive thoughts, I don't have time to solve problems"). This, of course, is a smaller bonus than solving the problem (that's why it is called "secondary benefit" in psychology), but you can still get some psychological dividends from people. For example, sympathy and condescension of others (if you are struggling and tearing forward - do not expect sympathy). And also, indulging in suffering, you can not think how to solve the problem, get out of trouble, to achieve success. Who will dare to demand constructive actions from a miserable person who is suffering so much!

For example, a person is constantly running through his head: "I don't want this life, it's better to die. There's no hope." Bonuses? Since there is no hope, it means there is no need to dash around trying to achieve something - you can sit and feel sorry for yourself, choosing the role of a martyr and victim. There's no need to set new goals and strive for them with a lot of effort, no need to become an adult and take RESPONSIBILITY for your own life. You can just exist, go with the flow, and see if it takes you out!

If you are overcome by intrusive thoughts, think about what indirect benefit you derive from them, what bonuses make you so happy that you can not give them up even under the threat of mental illness.

And take responsibility for your life. This is a very effective remedy for obsessive thoughts and a great stimulant for optimizing brain function, proven.

USED LITERATURE

1. Vartanyan I. A., Egorov V. Ya. Neurophysiology : textbook. SPb. Institute of Special Pedagogy and Psychology, 2014. 64 c.
2. Gaivoronsky I. V., Nichiporuk G. I., Gaivoronsky A. I. Anatomy of the central nervous system and sensory organs : textbook for bachelors. M. Yurait Publishing House, 2015. 293 c.
3. Nicholls J., Martin A., Vallas B., Fuchs P. From neuron to brain. M. Unitorial Urss, 2003. 672 c.
4. Agajanian N. A., Smirnov V. M. Normal physiology : textbook for medical universities. M. Med. inform. agency, 2007. 519 c.
5. Batuev A. C. Physiology of higher nervous activity and sensory systems : textbook for universities. 3rd edition, revised and supplemented. St. Petersburg. Peter, 2008. 317 c.
6. Blum F., Leiserson A., Hofstedter L. Brain, mind, behavior. M. : Mir, 1988. 248 c.
7. Dubynin V. A. A., Kamensky A. A., Sapin M. R., Sivoglazov V. I. Regulatory systems of the human body : textbook. R., Sivoglazov V. I. Regulatory systems of the human organism : textbook. M. Drofa, 2003. 368 c.
8. Shulgovsky V. V. Fundamentals of neurophysiology : textbook for universities. M. Aspect Press, 2000. 277 c.
9. Kuraev T. A., Aleynikova T. V., Dumbai V. N., Feldman G. L. N., Feldman G. L. Physiology of the central nervous system : textbook for universities. 2nd edition, revised and supplemented. Rostov n/D: Phoenix, 2000. 384 c.

10. Smirnov V. M., Yakovlev V. N. Physiology of the central nervous system : textbook. N. Physiology of the central nervous system : textbook. M. Academy, 2002. 352 c.
11. Uryvaev Yu. V., Dontsov R. G. Physiology of the central nervous system. The beginning of system integration. M. Aspect Press, 2007. 160 c.
12. Vartanyan I. A. Fundamentals of neurophysiology and higher nervous activity / I. A. Vartanyan. - SPb., 2003.
13. Vartanyan I. A. Neurophysiology : textbook / I. A. Vartanyan, V. Y. Egorov. - SPb. Institute of Special Pedagogy and Psychology, 2014.
14. Vartanyan I. A. Methodical recommendations for independent studies in neurophysiology, physiology of sensory systems and higher nervous activity / I. A. Vartanyan. - SPb. ISPiP, 2009.
15. Vartanyan I. A. Psychophysiology and higher nervous activity: word.-sprav. - SPb. : ISPiP, 2006. - [Electronic resource]. URL: http:// www.iprbookshop.ru/29993.html. - EBS "IPRbooks" (date of reference: 19.11.2018).
16. Egorov V. Ya. Methodological recommendations for independent work on anatomy: in 2 parts. Ch. 2 / V. Y. Egorov. - SPb. ISPiP, 2012.
17. Khomutov A. E. Physiology of the central nervous system / A. E. Khomutov. - Rostov/D: Phoenix, 2006.
18. Shulgovsky V. V. V. Fundamentals of neurophysiology / V. V. Shulgovsky. - M.: IP RAN, 1998.
19. Abramets I. Depressive syndrome: neurochemistry, neurophysiology, pharmacology / I. Abramets, D. Evdokimov, A. Talalaenko. - Palmarium Academic Publishing, 2014.

20. Agajanian N. A. Normal physiology : textbook for medical universities / N. A. Agajanian, V. M. Smirnov. - Moscow : Med. inform. agency, 2007.
21. Blum F. Brain, mind, behavior / F. Blum, A. Leiserson, L. Hofstadter. - M. : Mir, 1988.
22. Dubynin V. A. Dubynin. A. Regulatory systems of the human body : textbook / V. A. Dubynin, A. A. Kamensky, M. R. Sapin, V. I. Sivoglazov. - Moscow: Drofa, 2003.
23. Dykhan L. B. Introduction to the anatomy of the central nervous system : textbook / Dykhan L. B. ; Yuzh Feder. un. - Taganrog: Izd-vo Yuzh Feder. un-ta, 2016
24. Niyazmetov B., Akhmedov R., Mirzaolimov M. UNCOU'LED RESPIRATION IN BIRD MITICHONDRIA: CONNECTION WITH THERMOGENESIS //Bulletin of Namangan State University. - 2019. - T. 2019. - C. 18.
25. Mirzaolimov M. M., Abdullaev G. R., Abdullayev S. S. ROLE OF A CALORIE-RESTRICTED DIET IN PROLONGING THE LIFESPAN OF AN ORGANISM AND ITS MITOCHONDRIAL MECHANISMS //Scientific Bulletin of Namangan State University. - 2019. - T. 1. - №. 10. - C. 106-112.
26. Axmerov R. N. et al. ON THE POSSIBILITY OF UNCOUPLED MITOCHONDRIA IN BROWN FAT OF NEWBORN GUINEA PIGS //Scientific Bulletin of Namangan State University. - 2019. - T. 1. - №. 9. - C. 49-55.
27. Mirzaolimov M. M. et al. THE METHOD OF SEPARATION OF MITOCHONDRIAS AND DETERMINATION OF PHYSIOLOGICAL AND BIOCHEMICAL CHANGES IN

ORGANISMS IN ONTOGENESIS //Scientific Bulletin of Namangan State University. - 2020. - T. 2. - №. 3. - C. 175-178.

28. Soliev N., Mirzaolimov M. ACTION OF CALCIUM ON THE CONTENT OF PHOSPHOTYDYLCHOLIN, PHOSPHATYL ETHANOLAMINE AND THEIR LYSOFORMS IN THE RAT LIVER MITOCHONDRIA //Scientific Bulletin of Namangan State University. - 2019. - T. 1. - №. 3. - C. 69-71.
29. Mirzavalievich M. M., Adashaliyevich N. Q.. Soy protein, isoflavones, and cardiovascular health //International Journal of Scientific Trends. - 2023. - T. 2. - №. 4. - C. 10-18.
30. Mirzavalievich M. M., Adashaliyevich N. Q.. Soy protein, isoflavones, and cardiovascular health //International Journal of Scientific Trends. - 2023. - T. 2. - №. 4. - C. 10-18.
31. Anvarov F. R., Mirzaolimov M. M. HERIATRY (DEFINITION OF DEFINITIONS; CHALLENGES BEFORE THIS SCIENCE; SECTIONS AND ACHIEVEMENTS) //TA'LIM VA RIVOJLANISH TAHLILILI ONLAYN ILMIY JURNALI. - 2022. - C. 326-332.
32. Mirzaolimov M. M. Rakhimjonovich MAG IMPLICATION OF LIPID PEROXIDE Oxidation ON MITOCHONDRY OF RAT HEAD IN POSTNATAL ONTOGENESIS //INTERNATIONAL JOURNAL OF DISCOURSE ON INNOVATION, INTEGRATION AND EDUCATION. - 2020. - T. 1. - №. 5. - C. 78-86.
33. Sopiyev Sh. K., Ruziboeva S. I., Mirzaolimov M. M. Approaches to the development of children's and youth sports and the design of multi-year training and the design of multi-year training //PEDAGOGS jurnali. - 2022. - T. 5. - №. 1. - C. 162-170.

34. Mirzaolimov M. M. et al. GERIATRICS (DEFINITION OF CONCEPTS; TASKS FACING THIS SCIENCE; SECTIONS AND ACHIEVEMENTS) //TA'LIM VA RIVOJLANISH TAHLILI ONLAYN ILMIY JURNALI. - 2022. - C. 143-149.
35. Tadjibaeva G. I., Mirzaolimov M. M. KALAMUSHLARDA SURUNKALI EMOTIONAL STRESS MODELINI YARATISH VA ORGANIZMDAGI BIOKIMYOVI YETAKCHI TADQIQIQOTLAR ONLAYN ILMIY JURNALI //BARQARORLIK VA YETAKCHI TADQIQOTLAR ONLAYN ILMIY JURNALI. - 2022. - C. 166-170.
36. Atakhanova S. D. STRESS TAJSIRIDA ORGANISM A'ZO VA TIZIMLARIDAGI PHYSIOLOGIC-BIOKIMIY UZGARISHLAR //BARQARORLIK VA YETAKCHI TADQIQOTLAR ONLAYN ILMIY JURNALI. - 2022. - T. 2. - №. 4. - C. 215-219.
37. Sopiyev Sh. K., Ruziboeva S. I., Mirzaolimov M. M. Unity and Interrelationship between the structure of competition activity and the structure of training //PEDAGOGS journalali. - 2022. - T. 10. - №. 4. - C. 220-228.
38. Tadjibaeva G. I., Mirzaolimov M. M. MITOHONDRIANING TARKIBIY Tuzilishi VA BAJARADIGAN VAZIFALARI //PEDAGOGS jurnali. - 2022. - T. 8. - №. 1.
39. Mirzaolimov M. M. et al. Oxygen VA Nitrogen FAOL SHAKLLARI SYNTHESIS //PEDAGOGS jurnali. - 2022. - T. 7. - №. 1. - C. 394-400.
40. Mirzaolimov M. M. Abdullaev Gofurjon Rakhimjonovich, Effect of lipid peroxidation on rat liver mitochondria during postnatal ontogenesis //International journal of discourse on innovation. Integration and education, Uzb, December. - VOL. 11. - P. 78-86.

41.Karimov M. A. et al. RESULTS OF COMPLEX TREATMENT OF ORAL HEMANGIOMAS. - 2023.
42.Mavlanova S. MM Mirzaolimov Odam va hayvonlar fiziologiyasi (o 'quv-uslubiy majmua). Namangan, 2021.
43.Rakhmatullaev A. M., Mirzaolimov M. M. ROLE OF NA-K-NASOS. ACTIVE TRANSPORT OF CALCIUM AND WATER IONS IN CELL //Journal of new century innovations. - 2023. - T. 43. - №. 1. - C. 8-13.
44.Niyozov Q. A., Mirzaolimov M. M., qizi Kabirova Z. A. WOGONIN ON THE MECHANISM OF INFLUENZA VIRUS INFECTION OF ALVEOLAR MACROPHAGE INFLAMMATORY SUBSTANCES //Educational Research in Universal Sciences. - 2023. - T. 2. - №. 5. - C. 347-352.
45.Jalolov I. et al. FLAVONOIDS OF PAPAVER ANGRENICUM PLANT //Science and innovation. - 2023. - T. 2. - №. D10. - C. 91-94.
46.Mirzaolimov M. M. et al. ZHISMONIY RIVOZHLANISH, TAYORGARLIK DARAZASI VA INTELLECTUAL KOBILIYAT URTASIDAGI BOGLIKLIK // EDUCATION SCIENCE AND INNOVATIVE IDEAS IN THE WORLD. - 2024. - T. 43. - №. 7. - C. 58-63.
47.Mirzaolimov M. M. et al. BOSHLANFICH SINF Ў QUVCHILARINING ZHISMONIY RIVOZHLANISH DARAJASINI ANI KLASH USLUBLARI //World scientific research journal. - 2024. - T. 26. - №. 2. - C. 7-12.
48.Umirzaqovna X. N. N., Mirzavaliyevich M. M. FLAVOSANNI JIGAR MITOXONDRIYALARINING NAFAS OLISHI VA

OKSIDLANISHLI FOSFORLANISHIGA TA'SIRI // Journal of new century innovations. - 2023. - T. 43. - №. 1. - C. 21-27.
49. Mirzaolimov M. M., Niyozov K. A. FLAVANOID CONSERVING PLANT EXTRACTS //Educational Research in Universal Sciences. - 2023. - T. 2. - №. 6. - C. 187-193.
50. Mirzavaliyevich M. M., Xayitaliyevich Y. U. APIGENIN //" CANADA" INTERNATİONAL CONFERENCE ON DEVELOPMENTS İN EDUCATİON, SCİENCESAND HUMANİTİES. - 2023. - T. 9. - №. 1.
51. Mirzaolimov M. M. et al. EFFECTS OF NATURAL FLAVONES ON MEMBRANE PROPERTIES AND CITOTOXICITY OF HELA CELLS //MAGISTRLAR. - 2023. - C. 26.

MIX
Papier aus verantwortungsvollen Quellen
Paper from responsible sources
FSC® C105338

Printed by Books on Demand GmbH, Norderstedt / Germany